Lecture Notes in Mathematics

A collection of informal reports and seminars
Edited by A. Dold, Heidelberg and B. Eckmann, Zürich

34

Glen E. Bredon

University of California, Berkeley

Equivariant Cohomology Theories

1967

Springer-Verlag · Berlin · Heidelberg · New York

Library of Congress Catalog Card Number 67 - 25284 Title No. 7354.

Preface

These notes constitute the lecture notes to a series of lectures which the author gave at Berkeley in the spring of 1966.

Our central objective is to provide machinery for the study of the set $[[X;Y]]$ of equivariant homotopy classes of equivariant maps from the G-space X to the G-space Y (with base points fixed by G). (For various reasons we restrict our attention to the case in which G is a finite group.) An important tool for this study is equivariant cohomology theory. It is immediately seen, however, that the classical equivariant cohomology theory is quite inadequate for the task.

Our first object then is to develop an "equivariant classical cohomology theory" (as opposed to "classical equivariant cohomology theory") which is readily computable and which, for example, allows the development of an equivariant obstruction theory. This is done in Chapter I and the obstruction theory is considered in Chapter II. Our cohomology theory includes the classical theory as a special case.

An approximation to $[[X;Y]]$ is the stable object $\lim[[S^nX;S^nY]]$ which forms a group. If Y is a sphere, with a given G-action, this leads to the stable equivariant cohomotopy groups of a G-space X. These form an "equivariant generalized cohomology theory" and such theories are considered briefly in Chapter IV and related to the equivariant classical cohomology.

When X and Y are both spheres with (standard) involutions on them, the groups $\lim[[S^nX;S^nY]]$ are analogues of the stable homotopy groups of spheres and constitute the case of greatest interest to us at present. It is in fact this case which inspired the general theory expounded in these notes. Originally we intended to include a fifth chapter in these notes which would apply the general theory to this special case. However, the special case has since expanded in length and in importance to the extent that we have decided to publish our results on this topic separately. An outline of these results has appeared in our research announcement "Equivariant stable stems" in Bull. Amer. Math. Soc. 73 (1967) 269-273.

The main results in the present notes have been announced in "Equivariant cohomology theories," Bull. Amer. Math. Soc. 73 (1967) 266-268. Although we have restricted our attention, in these notes, to the case of finite groups it will be apparent that the theory goes through for cellular actions of discrete groups and this fact was incorporated in our research announcement (loc. cit.).

Those sections of the notes which contain relatively inessential material are marked with an asterisk. During the work on this subject the author was partially supported by the National Science Foundation grant GP-3990 and by a fellowship from the Alfred P. Sloan Foundation.

CONTENTS

Chapter IV. **Generalized Equivariant Cohomology**

Chapter I. Equivariant Classical Cohomology

1. G-complexes

Let G be a finite group. By a G-complex we mean a CW complex K together with a given action of G on K by cellular maps such that

(*) For each $g \in G$, $\{x \in K | g(x) = x\}$ is a subcomplex of K.

Note that for each $g \in G$, the fact that $g: K \to K$ and $g^{-1}: K \to K$ are assumed to be cellular implies that, in fact, each $g: K \to K$ in an automorphism of the given CW structure of K. Also it follows from the condition (*) that if $g \in G$ leaves any point $x \in K$ fixed then g must leave $K(x)$ pointwise fixed. ($K(A)$, for any subset $A \subset K$, denotes the smallest subcomplex of K containing A. It is a finite subcomplex iff A has compact closure.)

Let K be a G-complex and L a subcomplex invariant under G. Then an easy inductive argument on the skeletons of K shows that K has the equivariant homotopy extension property with respect to L. That is, if $f: K \to X$ is an equivariant map into any space X with a given G-action and if $F': L\times I \to X$ is any equivariant homotopy then there exists an equivariant homotopy $F: K\times I \to X$ extending F'.

Taking the case in which $X = L\times I \cup K\times\{0\}$ with f and F' the obvious maps we obtain the fact that $L\times I \cup K\times\{0\}$ is an equivariant retract of $K\times I$, the retraction being $F: K\times I \to L\times I \cup K\times\{0\}$. Let $B \subset X$ be the set of points x such that $F(x,1) \in L\times I$. Then B is a neighborhood of L in K and the composition

$$B\times I \xrightarrow{F} L\times I \cup K\times\{0\} \to K$$

is an equivariant strong deformation retraction of B onto L.

Now apply these facts to the G-complex $K\times I$ and the subcomplex $A = L\times I \cup K\times\{0\}$. Let U be a neighborhood of A possessing an equivariant strong deformation retraction onto A. Let $f\colon K \to I$ be a continuous function such that $f(x) = 0$ on some neighborhood of L and $f(x) = 1$ unless $x\times I \subset U$. By taking $x \to \inf\{f(g(x)) \mid g\in G\}$ we can assume that $f(g(x)) = f(x)$ for all $g\in G$. Define

$$F_t\colon K\times I \to K\times I$$

by $F_t(x,s) = (x,s(1-tf(x)))$. This forms a deformation of $K\times I$ into U which is equivariant and leaves A stationary. Following this by the deformation of U into A we see that $A = L\times I \cup K\times\{0\}$ is an equivariant strong deformation retract of $K\times I$.

Now identify $L\times\{1\}$ to a point, so that $K\times I$ becomes the mapping cylinder $M = K\times I/L\times\{1\}$ of the collapsing map $K \to K/L$. Now our deformation becomes a deformation retraction of M onto $K\times\{0\} \cup L\times I/L\times\{1\} \approx K\cup C_L$ (K with the cone C_L on L attached). On the other hand M can be deformed equivariantly into the face $K\times\{1\}/L\times\{1\} \approx K/L$. This shows that for any pair (K,L) of G-complexes, the G-complex K/L is of the same equivariant homotopy type as $K\cup C_L$.

Let us recall a construction central to the cohomology theory of CW complexes. Let K be a CW complex and pick an orientation for each cell of K. (If K is a G-complex it may be assumed that the operations of G preserve these orientations, because of (*), but this is not important.) Let $C_n(K)$ be the free abelian group generated by the n-cells of K. $C_n(K)$ is isomorphic to the singular homology group $H_n(K^n/K^{n-1};Z)$, or to $H_n(K^n, K^{n-1};Z)$.

Suppose that σ is an n-cell of K and let $f_\sigma: S^{n-1} \to K^{n-1}$ be a characteristic (attaching) map for σ. Collapsing K^{n-2} to a point, we obtain an induced map

$$S^{n-1} \to K^{n-1} \to K^{n-1}/K^{n-2} = \bigvee \tau/\dot{\tau} \tag{1.1}$$

where τ ranges over the (n-1)-cells of K ($\tau/\dot{\tau}$ is an oriented (n-1)-sphere and $\bigvee$ denotes the one point union). For each τ there is a projection $\bigvee\tau/\dot{\tau} \to \tau/\dot{\tau}$ (collapsing all other spheres). Let f_σ^τ denote the composed map

$$f_\sigma^\tau: S^{n-1} \to \tau/\dot{\tau}$$

The map (1.1) provides a singular homology class

$$\partial\sigma \in C_{n-1}(K) = H_{n-1}(K^{n-1}/K^{n-2})$$

and we clearly have that

$$\partial\sigma = \textstyle\sum_\tau [\tau:\sigma]\tau$$

where $[\tau:\sigma] = 0$ unless τ is an (n-1)-cell and, for an (n-1)-cell τ in K,

$$[\tau:\sigma] = \deg f_\sigma^\tau: S^{n-1} \to \tau/\dot{\tau}$$

(for fixed σ this is non-zero for only a finite number of cells τ, in fact f_σ^τ is a trivial map except for a finite number of cells τ). The correspondence $\sigma \to \partial\sigma$ generates a homomorphism

$$\partial: C_n(K) \to C_{n-1}(K)$$

which, in fact, is just the singular homology connecting homomorphism of the triple K^n, K^{n-1}, K^{n-2}. That is, ∂ is equivalent to the composition

$$H_n(K^n,K^{n-1}) \xrightarrow{\partial_*} H_{n-1}(K^{n-1}) \xrightarrow{j_*} H_{n-1}(K^{n-1},K^{n-2}).$$

We have that $\partial^2 = 0$ since the composition

$$H_{n-1}(K^{n-1}) \xrightarrow{j_*} H_{n-1}(K^{n-1},K^{n-2}) \xrightarrow{\partial_*} H_{n-2}(K^{n-2})$$

(part of the homology sequence of the pair (K^{n-1},K^{n-2})) is zero. Note that $\partial^2 = 0$ is equivalent to the equation

$$\sum_{\tau}[\omega:\tau][\tau:\sigma] = 0 \text{ for given } \omega,\sigma.$$

2. Equivariant cohomology theories

Let G be a finite group and let $\mathcal{G}$ denote the category of G-complexes and (continuous) equivariant maps. Let $\mathcal{G}_0$ denote the category of G-complexes with base point and base point preserving equivariant maps (base points are always assumed to be left fixed by each element of G and, in the case of G-complexes, to be a vertex). Let $\mathcal{G}^2$ be the category of pairs (K,L), $L \subset K$ a subcomplex, of G-complexes.

We use the abbreviation "Abel" to stand for the category of abelian groups.

An equivariant (generalized) cohomology theory on the category $\mathcal{G}$ is a sequence of contravariant functors

$$\mathcal{H}^n: \mathcal{G}^2 \to \text{Abel} \qquad (n \in Z)$$

together with natural transformations

$$\delta^n: \mathcal{H}^n(L,\emptyset) \to \mathcal{H}^{n+1}(K,L),$$

such that the following three axioms are satisfied (we put $\mathcal{H}^n(L) = \mathcal{H}^n(L,\emptyset)$):

(1) If f_0, f_1 are equivariantly homotopic maps (in $\mathcal{G}^2$) then $\mathcal{H}^n(f_0) = \mathcal{H}^n(f_1)$.

(2) The inclusion $(K,K\cap L) \subset (K\cup L,L)$ induces an isomorphism

$$\mathcal{H}^n(K\cup L,L) \xrightarrow{\approx} \mathcal{H}^n(K,K\cap L)$$

(3) If $(K,L) \in \mathcal{G}^2$ then the sequence

$$\cdots \to \mathcal{H}^n(K,L) \xrightarrow{j^*} \mathcal{H}^n(K) \xrightarrow{i^*} \mathcal{H}^n(L) \xrightarrow{\delta^*} \mathcal{H}^{n+1}(K,L) \to \cdots$$

is exact.

Remark. If G is abelian then the operations by elements of G are morphisms $\mathcal{G} \to \mathcal{G}$ (i.e. they are equivariant). Thus, in this case, each $\mathcal{H}^n(K,L)$ has a natural G-module structure.

There are functors $\mathcal{G}^2 \to \mathcal{G}_0$ and $\mathcal{G}_0 \to \mathcal{G}^2$ defined by $(K,L) \to K/L$ and $K \to (K,x_0)$ where x_0 is the base point of K, L/L is the base point of K/L (taken to be a disjoint point if $L = \emptyset$, in which case K^+ denotes $K/\emptyset$). Standard arguments can be used to translate the above axioms into an equivalent set of axioms for a "single space" theory on $\mathcal{G}_0$. (See, for example, G. W. Whitehead, Generalized homology theories, Trans. A. M. S. 102 (1962), pp. 227-283.)

In fact for $K \in \mathcal{G}_0$ let $SK = S\wedge K$ (with the obvious G action, trivial on the "circle factor" S) denote the reduced suspension of X. Then an equivariant cohomology theory on $\mathcal{G}_0$ is a sequence of contravariant functors

$$\tilde{\mathcal{H}}^n\colon \mathcal{G}_0 \to \text{Abel}$$

together with a sequence of natural transformations of functors σ^n

$$\sigma^n(K)\colon \tilde{\mathcal{H}}^n(K) \to \tilde{\mathcal{H}}^{n+1}(SK)$$

satisfying the following three axioms

(1') If f_0, f_1 are equivariantly homotopic (in $\mathcal{G}_0$) then $\tilde{\mathcal{H}}^n(f_0) = \tilde{\mathcal{H}}^n(f_1)$.

(2') $\sigma^n(K)$ is an isomorphism for each n and K.

(3') The sequence

$$\tilde{\mathcal{H}}^n(K/L) \to \tilde{\mathcal{H}}^n(K) \to \tilde{\mathcal{H}}^n(L)$$

is exact.

Most of the material of Chapter I of Eilenberg-Steenrod goes over directly to these generalized theories. Later on in these notes we shall show how to construct such theories using rather standard methods and shall consider some special cases of interest. We shall not concern ourselves with these matters at present, but shall confine ourselves to a discussion of "coefficient groups".

In non-equivariant theories the "coefficients" of the theory are defined to be $\mathcal{H}^*(pt)$ (or $\tilde{\mathcal{H}}^*(pt^+)$) and these (graded) groups are the primary distinguishing feature between different cohomology theories. In fact for (non-equivariant) "classical" theory (= cohomology theory + dimension axiom) the knowledge of the coefficient group ($\mathcal{H}^0(pt)$ in this case) allows <u>computation</u> of the cohomology of <u>any</u> finite simplicial complex. Essentially this is true because homotopy points (i.e. contractible objects such as simplexes) form the basic building blocks of all complexes.

For equivariant theory the situation is slightly more complicated, for now the "building blocks" are essentially the orbits (in an appropriate sense) of G. That is, the coset spaces G/H, where H ranges over the subgroups of G (not necessarily normal), form a representative set of building blocks.

Thus a "coefficient system" should contain all the groups $\mathcal{H}^*(G/H)$ (or $\tilde{\mathcal{H}}^*((G/H)^+))$). But this is not enough, for we must specify how the building blocks "fit together". That is, we must consider the equivariant maps $G/H \to G/K$ and a "coefficient system" must incorporate the induced homomorphisms

$$\mathcal{H}^*(G/K) \to \mathcal{H}^*(G/H)$$

in its structure.

In the following sections we define precisely what we mean by a coefficient system.

Terminology: A cohomology theory on $\mathcal{G}$ or $\mathcal{G}_0$ will be called "classical" (= "equivariant classical cohomology" but ≠ "classical equivariant cohomology" as defined, for example, in Steenrod and Epstein, Cohomology Operations) if it satisfies the additional "dimension" axiom:

(4) $\mathcal{H}^n(G/H) = 0$ for $n \neq 0$ and all H,

or, for a single space theory,

(4') $\tilde{\mathcal{H}}^n((G/H)^+) = 0$ for $n \neq 0$ and all H.

Later on, we shall prove existence and uniqueness theorems (of the Eilenberg-Steenrod type) for such "classical" theories.

3. The category of canonical orbits.

The category of canonical orbits of G, denoted by $\mathcal{O}_G$, is defined to be the category whose objects are the left coset spaces G/H and whose morphisms are the equivariant (with respect to left translation) maps $G/H \to G/K$.

For future reference we shall classify the equivariant maps $G/H \to G/K$. Suppose f is any map

$$f: G/H \to G/K$$

and put

$$f(H) = aK \quad \text{where } a \in G.$$

Then f is equivariant iff $f(gH) = gaK$ for all $g \in G$. Conversely, the formula $f(gH) = gaK$ defines a map (which must be equivariant) provided that

$$f(ghH) = f(gH)$$

for all $h \in H$. That is, we must have $ghaK = gaK$ for all $h \in H$. This is equivalent to $haK = aK$ and hence to

(3.1) $$a^{-1}Ha \subset K.$$

Thus we have the following result: Let $a \in G$ be such that $a^{-1}Ha \subset K$. Define

$$\hat{a}: G/H \to G/K$$

by

$$\hat{a}(gH) = gaK.$$

Then $\hat{a}$ is equivariant, that is, $\hat{a} \in \hom(G/H, G/K)$ and every equivariant map has this form. Also, clearly, $\hat{a} = \hat{b}$ iff $aK = bK$, that is, iff $a^{-1}b \in K$.

Suppose that (3.1) is satisfied. Then the inclusion $a^{-1}Ha \subset K$ induces a natural projection $G/a^{-1}Ha \to G/K$ (equivariant) and, similarly, the inclusion $H \subset aKa^{-1}$ induces $G/H \to G/aKa^{-1}$. Now <u>right</u> translation by a induces an equivariant map $R_a: G/H \to G/a^{-1}Ha$ (given by $gH \to gHa = ga(a^{-1}Ha)$) and also $R_a: G/aKa^{-1} \to G/K$. Clearly the diagram

(3.2)
$$\begin{array}{ccc} G/H & \longrightarrow & G/aKa^{-1} \\ {\scriptstyle R_a}\downarrow & \searrow^{\hat{a}} & \downarrow{\scriptstyle R_a} \\ G/a^{-1}Ha & \longrightarrow & G/K \end{array}$$

commutes. Thus equivariant maps are precisely those maps induced by inclusions of subgroups and by right translations.

In particular hom(G/H,G/H) consists of the right translations by elements of the normalizer N(H) of H (i.e. $a \in N(H)$ yields $gH \to gHa = gaH$). Since $R_a R_b = R_{ba}$, and generally $\hat{a}\hat{b} = \widehat{ba}$, the correspondence $a \to R_a^{-1}$ yields an isomorphism

(3.3) $$N(H)/H \approx \hom(G/H, G/H).$$

For example, let $G = Z_p$, where p is prime. Then $\mathcal{O}_G$ consists of the objects G/G and G/{e} (that is essentially of a point P and of G) together with the following morphisms

$$P \to P$$

$$G \to P$$

$$\hat{a}: G \to G \quad \text{for each } a \in G$$

(where here $\hat{a} = R_a$ takes g into ga).

4. Generic coefficient systems

(4.1) Definition. A (generic) coefficient system (for G) is defined to be a contravariant functor $\mathcal{O}_G \to$ Abel.

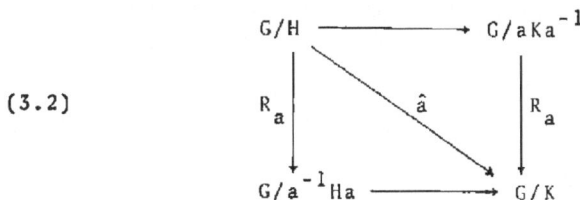

If M,N: $\mathcal{O}_G \to$ Abel are coefficient systems, a morphism T: M → N is a natural transformation of functors. With this definition, the (generic) coefficient systems for G form an abelian category $\mathcal{C}_G = \text{Dgram}(\mathcal{O}_G^*, \text{Abel})$. ($\mathcal{O}_G^*$ denotes the dual category to $\mathcal{O}_G$ and the fact that $\mathcal{C}_G$ is an abelian category is a special case of a result of Grothendieck; see Maclane, Homology, IX, 3.1, p. 258.)

Examples:

(1) Let $\mathcal{H}$ be an equivariant cohomology theory and let q be an integer. Define

$$h^q: \mathcal{O}_G \to \text{Abel}$$

by $h^q(G/H) = \mathcal{H}^q(G/H)$ and if $f: G/H \to G/K$ is equivariant, let $h^q(f) = \mathcal{H}^q(f): \mathcal{H}^q(G/K) \to \mathcal{H}^q(G/H)$.

(2) Let A be a G-module. Define

$$M: \mathcal{O}_G \to \text{Abel}$$

as follows: Let $M(G/H) = A^H$ (the set of stationary points of H in A). For $g \in G$ with $H \subset gKg^{-1}$ note that the operation by $g: A \to A$ takes A^K into A^H, (for $a \in A^K$ implies that $Hga \subset gKg^{-1}ga = gKa = ga$). Denote this map $A^K \to A^H$ by $g_{H,K}$. If $\hat{g} = \widehat{g'}$ so that $g^{-1}g' \in K$, then clearly $g_{H,K} = g'_{H,K}$. Thus, for $\hat{g}: G/H \to G/K$ we let

$$M(\hat{g}) = g_{H,K}: A^K \to A^H.$$

(3) Let Y be a G-space with a base point y_0. Define $\tilde{\omega}_q(Y) \in \mathcal{C}_G$, that is $\tilde{\omega}_q(Y): \mathcal{O}_G \to \text{Abel}$, as follows:

$$\tilde{\omega}_q(Y)(G/H) = \pi_q(Y^H, y_0)$$

$$\tilde{\omega}_q(Y)(\hat{g}) = g_\#: \pi_q(Y^K, y_0) \to \pi_q(Y^H, y_0)$$

where $g \in G$ satisfies $H \subset gKg^{-1}$, so that g maps $Y^K \to Y^H$ (see example 2). (In this example we assume each $\pi_1(Y^H, y_0)$ to be abelian when $q = 1$.)

Remark. Since $\hom(G/H,G/H) \approx N(H)/H$ we have that, for any coefficient system $M \in \mathcal{C}_G$, $M(G/H)$ possess a natural $N(H)/H$-module structure.

Let $M \in \mathcal{C}_G$. Since $\mathcal{O}_G$ contains, in particular, the objects $G = G/\{e\}$ and $P = G/G$ with the morphisms

$$1: P \to P$$

$$r: G \to P$$

$$\hat{a}: G \to G$$

we have that M "contains" the abelian groups $M(P)$ and $M(G)$ with the homomorphisms $M(1) = 1$ and

$$\varepsilon = M(r): M(P) \to M(G)$$

$$a_* = M(\hat{a}): M(G) \to M(G)$$

which satisfy $M(\widehat{ab}) = M(\hat{b}\hat{a}) = M(\hat{a})M(\hat{b})$ and $M(\hat{a})M(r) = M(r\hat{a}) = M(r)$; that is,

$$\begin{cases} (ab)_* = a_* b_* \\ a_* \varepsilon = \varepsilon . \end{cases}$$

Thus we may consider $M(G)$ to have a G-module structure defined by $(a,m) \to a_*(m)$ and $M(P)$ to have a trivial G-module structure and $\varepsilon : M(P) \to M(G)$ to be an equivariant homomorphism (i.e. $\varepsilon: M(P) \to M(G)^G$).

Of course, if $G = Z_p$ where p is prime, then this is all of the structure of an $M \in \mathcal{C}_G$. That is, in this case, a coefficient system consists of an abelian group M_0, an abelian group M_1 with a G-module structure and an homomorphism $\varepsilon : M_0 \to M_1^G$. Moreover, a morphism between two such systems M and M' is a commutative diagram of G-module homomorphisms:

$$\begin{array}{ccc} M_0 & \xrightarrow{\varepsilon} & M_1 \\ \downarrow & & \downarrow \\ M_0' & \xrightarrow{\varepsilon'} & M_1' \end{array}$$

For example, when $G = Z_p$ and Y is a G-space with base point, $\tilde{\omega}_q(Y)$ consists of the group $\pi_q(Y^G)$, the group $\pi_q(Y)$ on which G acts by the induced homomorphisms $g_\#: \pi_q(Y) \to \pi_q(Y)$, and the homomorphism $\varepsilon : \pi_q(Y^G) \to \pi_q(Y)^G \subset \pi_q(Y)$ induced by inclusion $Y^G \subset Y$.

5. Coefficient systems on a G-complex.

Let K be a G-complex. From K we form a category $\mathcal{K}$ whose objects are the finite subcomplexes of K and whose morphisms are as follows: If L and L' are finite subcomplexes of K, then $\hom(L,L')$ consists of all maps $g: L \to gL \subset L'$ for $g \in G$ ($\hom(L,L')$ may be empty). Note that we do not distinguish between maps induced by different elements of G if they are the same map.

Clearly the morphisms of $\mathcal{K}$ are just the inclusion maps $L \subset L'$, the maps $a: L \to aL$ induced by operations by elements of G, and the compositions of these.

We should note that for most purposes only the objects $K(\sigma)$ of $\mathcal{K}$ for cells σ of K are of importance, but for some constructions one needs the more general subcomplexes.

We define a canonical contravariant functor

$$\theta: \mathcal{K} \to \mathcal{O}_G$$

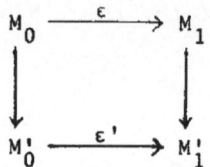

as follows: For $L \subset K$ a finite subcomplex, let $G_L =$ $\{g \in G | g$ leaves L pointwise fixed$\}$. We put

$$\theta(L) = G/G_L .$$

If $gL \subset L'$ and f denotes the map $L \to L'$ induced by operation by $g \in G$, then we see that

$$g^{-1}G_{L'}g \subset G_L$$

and we put $\theta(f) = \hat{g}: \theta(L') \to \theta(L)$, that is $\theta(f)$ is $\hat{g}: G/G_{L'} \to G/G_L$ which takes $g'G_{L'}$ into $g'gG_L$.

In other words, if $L \subset L'$ then $G_{L'} \subset G_L$ and θ(inclusion) is the natural map $G/G_L \to G/G_{L'}$ while if $g: L \to gL$ then $G_{gL} = gG_Lg^{-1}$ and $\theta(g: L \to gL): \theta(gL) = G/gG_Lg^{-1} \to G/G_L = \theta(L)$ is right multiplication by g.

Now if $M \in \mathcal{C}_G$ is a generic coefficient system, that is, if $M: \mathcal{O}_G \to \text{Abel}$ is a contravariant functor, then

$$M\theta: \mathcal{K} \to \text{Abel}$$

is a covariant functor and is called a (simple) coefficient system on K. We generalize this as follows:

A local coefficient system on K is a covariant functor

$$\mathcal{L}: \mathcal{K} \to \text{Abel}.$$

Again by Grothendieck's result, the local coefficient systems on K form an abelian category $\mathcal{LC}_K = \text{Dgram}(\mathcal{K}, \text{Abel})$.

The coefficient systems $M\theta: \mathcal{K} \to \text{Abel}$, for $M \in \mathcal{C}_G$, clearly form a subcategory $\mathcal{C}_K$ of $\mathcal{LC}_K$.

Notation. If $\mathcal{L} \in \mathcal{LC}_K$ and σ is a cell we let $\mathcal{L}(\sigma) = \mathcal{L}(K(\sigma))$ and for $K(\tau) \subset K(\sigma)$ we let $\mathcal{L}(\tau \to \sigma)$ denote $\mathcal{L}(\text{inclusion}: K(\tau) \to K(\sigma))$. Note that if $[\tau : \sigma] \neq 0$ then $K(\tau) \subset K(\sigma)$ so that $\tau \to \sigma$ is "in" $\mathcal{K}$.

6. Cohomology

Let $\mathcal{L}: \mathcal{K} \to$ Abel be in $\mathcal{L}\mathcal{C}_K$. Orient the cells of K in such a way that G preserves the orientations and define

$$C^q(K;\mathcal{L})$$

to be the group of all functions f on the q-cells of K with

$$f(\sigma) \in \mathcal{L}(\sigma).$$

Define $\delta : C^q(K;\mathcal{L}) \to C^{q+1}(K;\mathcal{L})$ by

$$(6.1) \qquad (\delta f)(\sigma) = \sum_{\tau} [\tau\colon \sigma]\mathcal{L}(\tau \to \sigma)f(\tau)$$

(which makes sense since $K(\tau) \subset K(\sigma)$ whenever $[\tau\colon \sigma] \neq 0$). In other words $(\delta f)(\sigma)$ is defined by "pushing" all coefficients to $\mathcal{L}(\sigma)$ and then taking the usual coboundary. This remark shows that $\delta\delta = 0$ since to compute $(\delta\delta f)(\omega)$ we push coefficients to $\mathcal{L}(\omega)$ and then compute (classical) coboundaries twice which necessarily gives zero. Of course, $\delta\delta = 0$ also follows by direct computation.

Now we define an operation of G on $C^q(K;\mathcal{L})$ as follows: If $g \in G$ and $f \in C^q(K;\mathcal{L})$ we put

$$(6.2) \qquad g(f)(\sigma) = \mathcal{L}(g)(f(g^{-1}\sigma)).$$

Here $\mathcal{L}(g)$ refers to $\mathcal{L}(g\colon K(g^{-1}\sigma) \to K(\sigma))$. Let us abbreviate

$$\mathcal{L}(g) = g_*.$$

Replacing σ by $g(\sigma)$ in (6.2) we obtain

$$(6.3) \qquad g(f)(g\sigma) = g_*(f(\sigma))$$

It is clear that the automorphism $f \to g(f)$ of $C^*(K;\mathcal{L})$ defines an <u>action</u> of G on $C^*(K;\mathcal{L})$ by chain mappings. Thus the fixed point set

$$C^q(K;\mathcal{L})^G = \{f \in C^q \mid g(f) = f \text{ for all } g \in G\}$$

is a subcomplex. It is also denoted by $C^q_G(K;\mathcal{L})$. By (6.3) $C^*(K;\mathcal{L})^G$ consists precisely of the <u>equivariant</u> cochains f (i.e. such that $f(g\sigma) = g_*(f(\sigma))$).

We define the equivariant cohomology group

(6.4) $$H^q_G(K;\mathcal{L}) = H^q(C^*(K;\mathcal{L})^G).$$

If $M \in \mathcal{C}_G$ (so that $M\theta \in \mathcal{C}_K \subset \mathcal{L}\mathcal{C}_K$) we use the abbreviation

(6.5) $$H^q_G(K;M) = H^q_G(K;M\theta).$$

If L is a subcomplex of K, invariant under G, then there is a restriction map $C^*(K;\mathcal{L}) \to C^*(L;\mathcal{L})$ whose kernel is the relative cochain group $C^*(K,L;\mathcal{L})$. There is a splitting homomorphism $C^*(L;\mathcal{L}) \to C^*(K;\mathcal{L})$ defined by extension of a cochain by zero (not a chain map). This clearly commutes with operations by G so that the sequence

$$0 \to C^*(K,L;\mathcal{L})^G \to C^*(K;\mathcal{L})^G \to C^*(L;\mathcal{L})^G \to 0$$

is exact. With the obvious definitions we obtain an induced cohomology exact sequence

$$\cdots \to H^n_G(K,L;\mathcal{L}) \to H^n_G(K;\mathcal{L}) \to H^n_G(L;\mathcal{L}) \to H^{n+1}_G(K,L;\mathcal{L}) \to \cdots$$

7.* <u>Equivariant maps</u>.

This section is not necessary to our main line of thought and it is included merely for the sake of completeness.

Let G and G' be finite groups and let $\varphi : G \to G'$ be a homomorphism. Let K be a G-complex, K' a G'-complex and let $\psi : K \to K'$ be a <u>cellular</u> map which is equivariant (i.e.

$\psi(g(x)) = \varphi(g)(\psi(x))$. The map ψ (together with φ) induces a functor

$$\Psi : \mathcal{K} \to \mathcal{K}'$$

(between the categories associated with K and K' respectively) as follows: If $L \subset K$, let $\Psi(L) = K'(\psi(L))$ and if f is the composition $L \xrightarrow{g} gL \subset L_1$ then $\Psi(f)$ is the obvious composition

$$K'(\psi(L)) \to \varphi(g)K'(\psi(L)) = K'(\varphi(g)\psi(L)) = K'(\psi(gL)) \subset K'(\psi(L_1)),$$

(By abuse of notation we might define Ψ on morphisms by writing $\Psi(g) = \varphi(g)$.)

Let $\mathcal{L}': \mathcal{K}' \to$ Abel be a local coefficient system on K'. Then $\mathcal{L}'\Psi: \mathcal{K} \to$ Abel is a local coefficient system on K. Suppose that $\mathcal{L}: \mathcal{K} \to$ Abel is any local coefficient system on K. Then we define a Ψ-<u>morphism</u> λ from $\mathcal{L}'$ to $\mathcal{L}$ to be a natural transformation

$$\lambda: \mathcal{L}'\Psi \to \mathcal{L}$$

of functors on $\mathcal{K}$. Now there is an obvious chain map $C^*(K;\mathcal{L}'\Psi) \to C^*(K;\mathcal{L})$ induced by λ and this is clearly equivariant with respect to the actions by G. Thus λ induces a homomorphism

(7.1) $$\lambda^*: H_G^*(K;\mathcal{L}'\Psi) \to H_G^*(K;\mathcal{L}).$$

We shall define a canonical homomorphism

(7.2) $$\Psi^*: H_{G'}^*(K';\mathcal{L}') \to H_G^*(K;\mathcal{L}'\Psi)$$

so that together with (7.1) we will obtain a homomorphism

$$\lambda^*\Psi^*: H_{G'}^*(K';\mathcal{L}') \to H_G^*(K;\mathcal{L})$$

(also denoted merely by λ^*).

In fact note that the cellularity of ψ implies that ψ induces a map $K^n/K^{n-1} \to K'^n/K'^{n-1}$ and hence induces a chain map

$$\psi_*: C_n(K) \to C_n(K').$$

Define

(7.3) $\Psi^*: C^*(K';\mathcal{L}') \to C^*(K;\mathcal{L}'\Psi)$

by

$$\Psi^*(f)(\sigma) = f(\psi_*(\sigma))$$

where the right hand side is shorthand for

$$\sum_\alpha n_\alpha \mathcal{L}'\big(K'(\tau_\alpha) \to K'(\psi(\sigma))\big) f(\tau_\alpha) \in \mathcal{L}'(K'(\psi(\sigma))) = \mathcal{L}'\Psi(\sigma)$$

where $\psi_*(\sigma) = \Sigma n_\alpha \tau_\alpha \in C_n(K')$.

Now we compute

$$\Psi^*(\varphi(g)(f))(\sigma) = (\varphi(g)(f))(\psi_*(\sigma)) = \mathcal{L}'(\varphi(g))(f(\varphi(g)^{-1}\psi_*(\sigma)))$$

$$= (\mathcal{L}'\Psi)(g)(f(\psi_*(g^{-1}\sigma))) = (\mathcal{L}'\Psi)(g)(\Psi^*(f)(g^{-1}\sigma))$$

$$= g(\Psi^*(f))(\sigma).$$

Thus, if $\varphi(g)(f) = f$ for all $g \in G$, then

$$g(\Psi^*(f)) = \Psi^*(\varphi(g)(f)) = \Psi^*(f).$$

Therefore (7.3) takes $C^*(K';\mathcal{L}')^{\varphi(G)}$ into $C^*(K;\mathcal{L}'\Psi)^G$. Since $C^*(K';\mathcal{L}')^{G'} \subset C^*(K';\mathcal{L}')^{\varphi(G)}$ we obtain a chain map $C^*(K';\mathcal{L}')^{G'} \to C^*(K;\mathcal{L}'\Psi)^G$ which induces our promised map (7.2) upon passage to homology.

The situation with <u>simple</u> coefficient systems is slightly more complicated, and we shall now discuss this case. We define a functor

$$\Phi : \mathcal{O}_G \to \mathcal{O}_{G'}$$

by putting $\Phi(G/H) = G'/\varphi(H)$ and, if $a^{-1}Ha \subset K$ as in (3.1), so that $\varphi(a)^{-1}\varphi(H)\varphi(a) \subset \varphi(K)$ we put $\Phi(\hat{a}: G/H \to G/K) = \widehat{\varphi(a)}: G'/\varphi(H) \to G'/\varphi(K)$.

The diagram

$$\begin{array}{ccc} \mathcal{K} & \xrightarrow{\Psi} & \mathcal{K}' \\ \downarrow{\scriptstyle\theta} & & \downarrow{\scriptstyle\theta'} \\ \mathcal{O}_G & \xrightarrow{\Phi} & \mathcal{O}_{G'} \end{array}$$

does not generally commute since

$$\theta'\Psi(L) = \theta'(K'(\psi(L))) = G'/G'_{\psi(L)}$$

while

$$\Phi\theta(L) = \Phi(G/G_L) = G'/\varphi(G_L)$$

and $\varphi(G_L) \subset G'_{\psi(L)}$ are not generally equal. However the projection

$$G'/\varphi(G_L) \to G'/G'_{\psi(L)}$$

is clearly functorial and provides a natural transformation

$$(7.4) \qquad \Phi\theta \to \theta'\Psi$$

of functors. Let $M' \in \mathcal{C}_{G'}$ be a generic coefficient system for G'. Since M' is a contravariant functor $\mathcal{O}_{G'} \to$ Abel, the transformation (7.4) induces a natural transformation

$$(7.5) \qquad M'\theta'\Psi \to M'\Phi\theta$$

of functors $\mathcal{K} \to$ Abel. In other words, (7.5) is a Ψ-morphism

$$(7.6) \qquad M'\theta' \to M'\Phi\theta.$$

Thus we have an induced homomorphism

$$(7.7) \qquad H^*_{G'}(K';M') \to H^*_G(K;M'\Phi)$$

(where the θ and θ' have been dropped in accordance with our notation conventions).

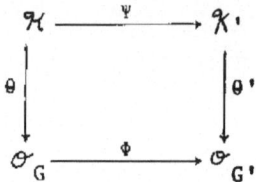

If $M \in \mathcal{C}_G$ and $M' \in \mathcal{C}_{G'}$ we define a φ-morphism $M' \to M$ to be a natural transformation

$$M'\Phi \to M$$

of functors $\mathcal{O}_G \to$ Abel. Clearly, in combination with (7.7), every φ-morphism $M' \to M$ induces a homomorphism

(7.8) $$H^*_{G'}(K';M') \to H^*_G(K;M).$$

8.* Products

Suppose that K is a G-complex and K' is a G'-complex. Then $K\times K'$ with the product cell-structure and the weak topology is a $G\times G'$-complex in the obvious way. If $\mathcal{L}$ and $\mathcal{L}'$ are local coefficient systems on K and K' respectively then define

$$\mathcal{L} \hat{\otimes} \mathcal{L}' \in \mathcal{L}\mathcal{C}_{K\times K'}$$

by $(\mathcal{L} \hat{\otimes} \mathcal{L}')(W) = \mathcal{L}(\pi_1 W) \otimes \mathcal{L}'(\pi_2 W)$ where $\pi_1 \colon K\times K' \to K$ and $\pi_2 \colon K\times K' \to K'$ are the projections. The definition of $\mathcal{L} \hat{\otimes} \mathcal{L}'$ on morphisms is obvious.

Suppose that $f \in C^p(K;\mathcal{L})$ and $f' \in C^q(K';\mathcal{L}')$. Define

$$f\times f' \in C^{p+q}(K\times K'; \mathcal{L} \hat{\otimes} \mathcal{L}')$$

by

$$(f\times f')(\sigma\times\tau) = f(\sigma) \otimes f'(\tau)$$

where σ and τ are (oriented) p and q-cells respectively ($f\times f'$ vanishes elsewhere). $(f,f') \to f\times f'$ is obviously bilinear.

If $g \in G$ and $g' \in G'$ then clearly

$$(g\times g')(f\times f') = g(f) \times g'(f').$$

It is also clear that $\delta(f\times f') = (\delta f)\times f' + (-1)^p f\times\delta f'$. Thus $\times$ induces a chain map

$$C^p_G(K;\mathcal{L}) \otimes C^q_{G'}(K';\mathcal{L}') \to C^{p+q}_{G\times G'}(K\times K';\mathcal{L}\hat{\otimes}\mathcal{L}')$$

and consequently, a "cross-product":

$$H^p_G(K;\mathcal{L}) \otimes H^q_{G'}(K';\mathcal{L}') \to H^{p+q}_{G\times G'}(K\times K';\mathcal{L}\hat{\otimes}\mathcal{L}').$$

If $\mathcal{L}$ and $\mathcal{L}'$ are simple then so is $\mathcal{L}\hat{\otimes}\mathcal{L}'$ as the reader can check.

An internal product, the "cup-product" can be derived from the cross-product by means of equivariant diagonal approximations. However, we have not given the necessary background for this since the definition of the cup product is more easily obtained as a consequence of general facts which we shall develop later in these notes.

9*. Another description of cochains.

We define an element

$$\underline{C}_n(K;Z) \in \mathcal{C}_G$$

by $\underline{C}_n(K;Z)(G/H) = C_n(K^H;Z)$ together with the obvious values on morphisms of $\mathcal{O}_G$. These objects, for $n = 0,1,2,\ldots$, form a chain complex in the abelian category $\mathcal{C}_G$. We can form the homology $\underline{H}_n(K;Z) = H_n(\underline{C}_*(K;Z)) \in \mathcal{C}_G$ of this chain complex. Clearly, this is just $\underline{H}_n(K;Z)(G/H) = H_n(K^H;Z)$ together, again, with the obvious values on morphisms. Similar considerations apply to the relative case.

Let $f \in C^n_G(K;M)$ where $M \in \mathcal{C}_G$. Then for an n-cell σ, $f(\sigma) \in M(G/G_\sigma)$. Suppose that $\sigma \in K^H$. Then $H \subset G_\sigma$ so that we have an element

$$M(G/H \to G/G_\sigma)f(\sigma) \in M(G/H).$$

Denote this element by $\hat{f}(G/H)(\sigma)$. This map clearly extends to a homomorphism

(9.1) $\hat{f}(G/H)\colon C_n(K^H;Z) \to M(G/H)$.

It is easily checked that (9.1) is natural with respect to the morphisms of $\mathcal{O}_G$, so that $\hat{f}\colon \underline{C}_n(K;Z) \to M$ is a natural transformation of functors. That is,

(9.2) $\hat{f} \in \mathrm{Hom}(\underline{C}_n(K;Z),M)$

where Hom refers to the morphisms of the abelian category $\mathcal{C}_G$. Conversely, suppose we are given an element $\hat{f} \in \mathrm{Hom}(\underline{C}_n(K;Z),M)$. Let σ be an n-cell of K and regard σ as an element of $C_n(K^{G_\sigma};Z)$. Define

$$f(\sigma) = \hat{f}(G/G_\sigma)(\sigma) \in M(G/G_\sigma)$$

so that $f \in C^n(K;M)$. Let us check that f is equivariant. Applying the fact that $\hat{f}$ is natural to the morphism $\hat{g}\colon G/G_{g\sigma} = G/gG_\sigma g^{-1} \to G/G_\sigma$ of $\mathcal{O}_G$, we see that the diagram

$$\begin{array}{ccc} C_n(K^{G_\sigma};Z) & \xrightarrow{\hat{f}(G/G_\sigma)} & M(G/G_\sigma) \\ \Big\downarrow g_* & & \Big\downarrow g_* = M(\hat{g}) \\ C_n(K^{G_{g\sigma}};Z) & \xrightarrow{\hat{f}(G/G_{g\sigma})} & M(G/G_{g\sigma}) \end{array}$$

commutes. Thus $f(g\sigma) = \hat{f}(G/G_{g\sigma})(g\sigma) = g_*(\hat{f}(G/G_\sigma)(\sigma)) = g_*(f(\sigma))$ as claimed.

We have demonstrated an isomorphism

(9.3) $C^n_G(K;M) \approx \mathrm{Hom}(\underline{C}_n(K;Z),M)$

given by $f \to \hat{f}$. It is clear that this isomorphism preserves the coboundary operators. Thus we may pass to homology and obtain the isomorphism

(9.4) $\quad H^n_G(K;M) \approx H^n(\mathrm{Hom}(\underline{C}_*(K;Z),M))$.

Since Hom is left exact on $\mathcal{C}_G$ we obtain a canonical homomorphism

(9.5) $\quad H^n_G(K;M) \to \mathrm{Hom}(\underline{H}_n(K;Z),M)$.

It is also easy to check that _if_ K _has no_ (n-1)-_cells_, so that $\underline{C}_{n-1}(K;Z) = 0$, then (9.5) is an _isomorphism_ (triviality of $\underline{H}_{n-1}(K;Z)$, or even of $\underline{H}_q(K;Z)$ for $0 < q < n$, is not sufficient for this).

Remark. If A is a G-module and $M \in \mathcal{C}_G$ is the corresponding coefficient system as defined in §4, example 2, then an equivariant homomorphism $C_n(K;Z) \to A$ must take $C_n(K^H;Z) \subset C_n(K;Z)^H$ into $A^H = M(G/H)$. Thus it is clear that we have an isomorphism

$$\mathrm{Hom}_{Z(G)}(C_n(K;Z),A) \approx \mathrm{Hom}(\underline{C}_n(K;Z),M) \approx C^n_G(K;M).$$

The left hand side is, by definition, the classical equivariant cochain group with coefficients in the G-module A.

10.* A spectral sequence.

We shall show that the abelian category $\mathcal{C}_G$ contains sufficiently many projectives and injectives. However, projective resolutions of length one (or even of finite length) do not generally exist, in contrast to the category Abel. Thus instead of a universal coefficient sequence linking homology and cohomology we obtain a _spectral_ sequence.

For a set S let F(S) denote the free abelian group based on S. Suppose that S is a G-set. Define an element

$$F_S \in \mathcal{C}_G \quad \text{by} \quad F_S(G/H) = F(S^H)$$

together with the obvious values on morphisms of $\mathcal{O}_G$ (see §4, example 2). For example, if S is the set of n-cells of a G-complex K which are not in the G-subcomplex L, then $F_S \approx \underline{C}_n(K,L;Z)$.

(10.1) Proposition. F_S is projective.

Proof. Let

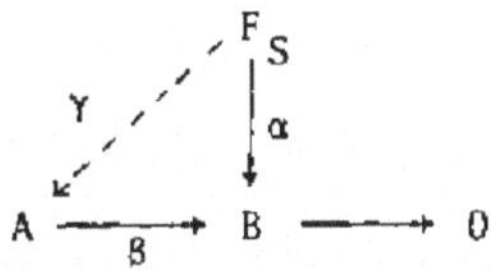

be a diagram in $\mathcal{C}_G$ with exact row and with γ to be constructed. Let $S' \subset S$ be a subset containing exactly one element from each orbit of G on S. Given $s \in S'$, consider s as an element of $F(S^{G_s}) = F_S(G/G_s)$. Then $\alpha(s) \in B(G/G_s)$. Define $\gamma(s) \in A(G/G_s)$ to be any element with $\beta(\gamma(s)) = \alpha(s)$. For $g \in G$ we let $\gamma(gs) = g_*\gamma(s) \in A(G/G_{gs})$ (where $g_* = A(\hat{g}: G/G_{gs} \to G/G_s)$). For $H \subset G_s$ let j denote the projection $G/H \to G/G_s$. The element s represents an element of $F(S^H) = F_S(G/H)$, namely $F_S(j)(s)$. We define $\gamma(F_S(j)(s)) = A(j)\gamma(s)$. Now γ has been defined on a set of free generators of $F_S(G/H)$ for every $H \subset G$. Thus there is a unique extension to $F_S(G/H)$ for all H. This extension is clearly a morphism $F_S \to A$ with $\beta\gamma = \alpha$, as claimed.

(10.2) Corollary. $C^n_G(K,L;M) \approx \mathrm{Hom}(\underline{C}_n(K,L;Z),M)$.

Proof. The exact sequence

$$0 \to \underline{C}_n(L;Z) \to \underline{C}_n(K;Z) \to \underline{C}_n(K,L;Z) \to 0$$

of projective objects in $\mathcal{C}_G$ induces an exact sequence via the functor Hom(· ,M) and the result follows.

(10.3) Corollary. $C^n_G(K,L;M)$ is an exact functor of M.

Proof. This is immediate from (10.2).

It follows from (10.3) that an exact sequence $0 \to M' \to M \to M'' \to 0$ in $\mathcal{C}_G$ induces a long exact cohomology sequence of (K,L).

At the end of this section we shall show that $\mathcal{C}_G$ contains sufficiently many projectives. In fact if S is the disjoint union of all of the G-sets G/H for $H \subset G$ then F_S is a (projective) generator of the category $\mathcal{C}_G$. Since $\mathcal{C}_G$ obviously satisfies Grothendieck's axiom AB5 (arbitrary direct sums and exactness of the direct limit functor) it follows by a result of Grothendieck that $\mathcal{C}_G$ possesses sufficiently many injectives (see Mitchell: Theory of Categories).

Let $M \in \mathcal{C}_G$ and let M^* be an injective resolution of M. Consider the double complex

$$\mathrm{Hom}(\underline{C}_*(K,L;Z),M^*).$$

Standard homological algebra applied to this double complex yields a spectral sequence with

(10.4) $$E_2^{p,q} = \mathrm{Ext}^p(\underline{H}_q(K,L;Z),M) \Longrightarrow H_G^{p+q}(K,L;M).$$

(This notation means that $E_r^{p,q}$ converges to $E_\infty^{p,q}$ which is the graded group associated with a filtration of $H_G^{p+q}(K,L;M)$. Also Ext^p refers to the pth right derived functor of Hom in the category $\mathcal{C}_G$.)

By way of illustration we shall compute $\mathrm{Ext}^p(A,M)$ in two rather elementary cases.

<u>Example 1</u>. Let $A \in \mathcal{C}_G$ be defined by $A(G) = Z$, with trivial G-operators, and $A(G/H) = 0$ for $H \neq \{e\}$. Let F_* be a $Z(G)$-free resolution of Z. Then $\underline{F}_*$, defined by $\underline{F}_*(G) = F_*$ and $\underline{F}_*(G/H) = 0$ for $H \neq \{e\}$, is a projective resolution of A in $\mathcal{C}_G$. Clearly $\mathrm{Hom}(\underline{F}_*;M) \approx \mathrm{Hom}_{Z(G)}(F_*;M(G))$ so that

$$\mathrm{Ext}^p(A,M) \approx H^p(G;M(G)),$$

where the right hand side is the classical cohomology of G with coefficients in the G-module M(G). If K is a connected G-complex on which G acts <u>freely</u> and such that

$$H_q(K;Z) = 0 \quad \text{for} \quad 0 < q < N$$

then in (10.4) we have $E_2^{p,q} \approx \delta_0^q \mathrm{Ext}^p(A,M) \approx \delta_0^q H^p(G;M(G))$ for $q < N$. Consequently, we have an isomorphism

$$H_G^n(K;M) \approx H^n(G;M(G)) \quad \text{for } n < N.$$

<u>Example 2</u>. Let B be an abelian group and let $\underline{B} \in \mathcal{C}_G$ be defined by $\underline{B}(G/H) = B$ and $\underline{B}(j) = 1$ for all morphisms j in $\mathcal{O}_G$. Then, if M^* is an injective resolution of M, we have

$$\mathrm{Hom}(\underline{B},M^*) \approx \mathrm{Hom}(\underline{B}(P),M^*(P)) = \mathrm{Hom}(B,M^*(P))$$

where P is the point G/G. $M^*(P)$ is clearly an injective resolution of M(P) in Abel. Hence

$$\mathrm{Ext}^p(\underline{B},M) \approx \mathrm{Ext}^p(B,M(P))$$

where the right hand side is Ext in Abel. That is

$$\begin{cases} \mathrm{Ext}^0(\underline{B},M) = \mathrm{Hom}(\underline{B},M) \approx \mathrm{Hom}(B,M(P)) \\ \mathrm{Ext}^1(\underline{B},M) \approx \mathrm{Ext}(B,M(P)) \\ \mathrm{Ext}^p(\underline{B},M) = 0 \quad \text{for} \quad p > 1. \end{cases}$$

In particular, if B is free abelian then $\mathrm{Ext}^p(\underline{B},M) = 0$ for $p > 0$, that is, $\underline{B}$ is projective in $\mathcal{C}_G$ if B is projective in Abel. (Of course, this also follows directly from (10.1) in the case in which G acts trivially on S.)

Let us return to the general discussion. There is an edge homomorphism

$$H_G^n(K,L;M) \to \mathrm{Hom}\,(\underline{H}_n(K,L;Z),M)$$

of (10.4) (coinciding with (9.5) when $L = \emptyset$). Clearly this is an <u>isomorphism</u> if each $\underline{H}_q(K,L;Z)$ is projective for $q < n$.

For example suppose that $n > 1$, that K possesses stationary points (e.g. k_0) and that

$$\tilde{\omega}_q(K,k_0) = 0 \text{ for } q < n.$$

The Hurewicz theorem, applied to each K^H, shows that the (obvious) Hurewicz homomorphism (in $\mathcal{C}_G$)

$$\tilde{\omega}_q(K,k_0) \to \underline{H}_q(K;Z)$$

is an isomorphism for $0 < q \leq n$. Thus

$$(10.5) \qquad H_G^n(K;M) \approx \mathrm{Hom}(\tilde{\omega}_n(K,k_0),M)$$

in this case.

We shall now justify our earlier contention that there are enough projectives in $\mathcal{C}_G$. For any G-sets S and T let E(S,T) denote the set of equivariant maps $S \to T$. For $K \subset G$, the assignment $f \to f(K)$ clearly yields a one-one correspondence

$$E(G/K,S) \xrightarrow{\approx} S^K.$$

(It is of interest to reconsider the material of §3 and the examples of §4 in this light.) Thus

$$F_{G/H}(G/K) = F((G/H)^K) = F(E(G/K,G/H)).$$

Now if $\alpha \in M(G/H)$ the map $f \to M(f)(\alpha)$ of

$$E(G/K,G/H) \to M(G/K)$$

induces a homomorphism $F(E(G/K,G/H)) \to M(G/K)$. This is clearly natural in G/K and hence is a morphism

$$\varphi_\alpha : F_{G/H} \to M$$

in $\mathcal{C}_G$. It is also clear that the generator $H/H \in F_{G/H}(G/H)$ corresponds to $1 \in E(G/H,G/H)$ and hence that φ_α maps it into $\alpha \in M(G/H)$.

We shall now explicitly exhibit a projective which maps onto a given $M \in \mathcal{C}_G$. For $\alpha \in M(G/H)$ let S_α be a copy of the G-set G/H and let $S(M) = \bigcup_\alpha S_\alpha$ be the disjoint union of these for all $\alpha \in M(G/H)$ and all $H \subset G$. Then $F_{S(M)} = \sum_\alpha F_{S_\alpha}$. The homomorphisms $\varphi_\alpha : F_{S_\alpha} \to M$ yield a homomorphism

$$\varphi = \sum \varphi_\alpha : F_{S(M)} \to M \qquad (10.6)$$

which is clearly surjective.

Chapter II. Equivariant Obstruction Theory

In this chapter we shall assume that the reader is reasonably familiar with obstruction theory on CW-complexes. We shall attempt to strike a reasonable balance between giving no details on the one hand and developing the theory from scratch on the other by making use of the results, without proof, of the classical theory.

1. The obstruction cocycle

In this section $n \geq 1$ will be an integer, fixed throughout the discussion. Let K be a G-complex and L a G-subcomplex. Let Y be a G-space. We shall assume, for simplicity, that the set Y^H of stationary points of H on Y is non-empty, arcwise connected and n-simple for each subgroup $H \subset G$, (We note here that the theory could be generalized to relative CW-complexes (K,L) with no trouble.)

Assume that we are given an equivariant map $\varphi: K^n \cup L \to Y$. Let σ be an (n+1)-cell of K and let $f_\sigma: S^n \to K^n$ be a characteristic map for σ (note that the characteristic maps may be chosen equivariantly).

The subgroup G_σ leaves $K(\sigma)$, and hence Im f_σ, stationary. It follows that G_σ leaves $\mathrm{Im}(\varphi \circ f_\sigma)$ stationary. That is,

$$(\varphi \circ f_\sigma)(S^n) \subset Y^{G_\sigma}.$$

Thus $\varphi \circ f_\sigma$ defines an element $c_\varphi(\sigma) \in \pi_n(Y^{G_\sigma})$, and clearly $c_\varphi(\sigma) = 0$ if σ is in L. But, with $\tilde{\omega}_n(Y)$ defined as in example (3) of Chap. I, §4, this defines a cochain

$$c_\varphi \in C^{n+1}(K,L;\tilde{\omega}_n(Y)),$$

Now $c_\varphi(g\sigma)$ is represented by $\varphi \circ f_{g\sigma}: S^n \to Y^{G_{g\sigma}} = Y^{gG_\sigma g^{-1}}$ and $\varphi \circ f_{g\sigma} = \varphi \circ g \circ f_\sigma = g \circ \varphi \circ f_\sigma$ so that $c_\varphi(g\sigma) = g_\#(c_\varphi(\sigma))$. This means that c_φ is an equivariant cochain (by the defintion of $\tilde{\omega}_n(Y)$), that is

$$c_\varphi \in C_G^{n+1}(K,L;\tilde{\omega}_n(Y)).$$

It is called the obstruction cochain.

(1.1) Proposition. $\delta c_\varphi = 0$.

Proof. Let τ be an $(n+2)$-cell and consider the computation of $(\delta c_\varphi)(\tau)$. To calculate this, one "pushes" the coefficients to those on τ; that is to $\pi_n(Y^{G_\tau})$, and calculates the classical coboundary. But c_φ restricted to $K(\tau)$ and with coefficients pushed to $\pi_n(Y^{G_\tau})$ is just the obstruction cochain, in the classical sense, to extending $\varphi | K^n \cap K(\tau)$ to $K^{n+1} \cap K(\tau)$. Thus $(\delta c_\varphi)(\tau) = 0$ is a fact from the classical theory.

(1.2) Proposition. $c_\varphi = 0$ iff φ can be extended equivariantly to $K^{n+1} \cup L$.

Proof. If $c_\varphi(\sigma) = 0$ then clearly we may extend φ to $K^n \cup L \cup \sigma$ in such a way that $\varphi(\sigma) \subset Y^{G_\sigma}$. Define, for $g \in G$ and $x \in \sigma$,

$$\varphi(gx) = g\varphi(x) \in g(Y^{G_\sigma}) = Y^{gG_\sigma g^{-1}} = Y^{G_{g\sigma}}.$$

If $gx = g'x$ then $g' = gh$ for some $h \in G_\sigma$ so that $g'\varphi(x) = g\varphi(x)$ (since $\varphi(x) \in Y^{G_\sigma}$), which shows that this definition is valid. The proof is completed by taking an $(n+1)$-cell from each orbit of G on the $(n+1)$-cells and following the procedure above.

Now suppose that φ and θ are equivariant maps $K^n \cup L \to Y$ and let $F: (K^{n-1} \cup L) \times I \to Y$ be an equivariant homotopy between $\varphi | K^{n-1} \cup L$ and $\theta | K^{n-1} \cup L$. Define an equivariant map $\varphi \#_F \theta$: $(K \times I)^n \cup (L \times I) \to Y$ by

$$\begin{cases} (\varphi \#_F \theta)(x,0) = \varphi(x) \\ (\varphi \#_F \theta)(x,1) = \theta(x) \\ (\varphi \#_F \theta)(x,t) = F(x,t). \end{cases}$$

If $\varphi | K^{n-1} \cup L = \theta | K^{n-1} \cup L$ and F is the constant homotopy $\#$ will denote $\#_F$.

The deformation cochain $d_{\varphi,F,\theta} \in C^n_G(K,L;\tilde{\omega}_n(Y))$ is defined by

$$d_{\varphi,F,\theta}(\sigma) = c_{\varphi \#_F \theta}(\sigma \times I).$$

It is clear that

(1.3) $$\delta d_{\varphi,F,\theta} = c_\theta - c_\varphi .$$

If $\#_F = \#$, that is if F is constant, then we put $d_{\varphi,\theta} = d_{\varphi,F,\theta}$.

(1.4) Proposition. *Let* $\varphi: K^n \cup L \to Y$ *be equivariant and let* $d \in C^n_G(K,L;\tilde{\omega}_n(Y))$. *Then there is an equivariant map* $\theta: K^n \cup L \to Y$, *coinciding with* φ *on* $K^{n-1} \cup L$, *such that* $d_{\varphi,\theta} = d$.

Proof. Let σ be an n-cell of K, not in L, and choose a characteristic map $f_\sigma: (B^n, S^{n-1}) \to (K^n, K^{n-1})$ for σ. Let $J^n = B^n \times \{0\} \cup S^{n-1} \times I \subset B^n \times I$ and define $\Psi: J^n \to Y^{G_\sigma}$ by $\Psi(x,t) = \varphi(f_\sigma(x))$. As shown in non-equivariant obstruction theory, Ψ may be extended to a map $\Psi': \partial(B^n \times I) \to Y^{G_\sigma}$ representing the element (or any element) $d(\sigma) \in \pi_n(Y^{G_\sigma})$. It is clear that such extensions may be chosen *equivariantly*, since d is an equivariant cochain.

Now θ can be defined by $\theta|K^{n-1}\cup L = \varphi|K^{n-1}\cup L$ and, for an n-cell σ and $x\in\sigma$,

$$\theta(x) = \Psi'(f_\sigma^{-1}(x),1).$$

It is clear that $d_{\varphi,\theta} = d$.

The cocycle $c_\varphi \in C_G^{n+1}(K,L;\tilde{\omega}_n(Y))$ represents a cohomology class

$$[c_\varphi]\in H_G^{n+1}(K,L;\tilde{\omega}_n(Y))$$

which depends, by (1,3), only on the <u>equivariant homotopy class</u> of $\varphi|K^{n-1}\cup L$. Moreover, if $[c_\varphi] = 0$, then by (1.4) $\varphi|K^{n-1}\cup L$ extends to θ: $K^n\cup L \to Y$ such that $c_\theta = 0$ (choose d with $\delta d = -c_\varphi$). Hence, by (1.2), we have the following result:

(1.5) <u>Theorem</u>. <u>Let</u> φ: $K^n\cup L \to Y$ <u>be equivariant. Then</u> $\varphi|K^{n-1}\cup L$ <u>can be extended to an equivariant map</u> $K^{n+1}\cup L \to Y$ <u>iff</u> $[c_\varphi] = 0$.

<u>Remark</u>. Suppose that φ,θ: $K \to Y$ are equivariant and that F: $(K^{n-1}\cup L)\times I \to Y$ is an equivariant homotopy between the restrictions of φ and θ to $K^{n-1}\cup L$. As above we obtain an equivariant map $\varphi\#_F\theta$= $(K^{n-1}\times I)\cup Q \to Y$ where $Q = (L\times I)\cup(K\times\partial I)$. Then the obstruction to extending $\varphi\#_F\theta$ to $(K^n\times I)\cup Q$ is

$$c_{\varphi\#_F\theta}\in C_G^{n+1}(K\times I, L\times I\cup K\times\partial I;\tilde{\omega}_n(Y)).$$

This group is isomorphic to $C_G^n(K,L;\tilde{\omega}_n(Y))$ and this isomorphism takes $c_{\varphi\#_F\theta}$ into $d_{\varphi,F,\theta}$ (now a cocycle).

2. Primary obstructions

At various points in this section we shall make one or more of the following assumptions:

(1) Y^H is r-simple, non-empty and arcwise connected for all r and $H \subset G$ (e.g. $\tilde{\omega}_0(Y) = 0 = \tilde{\omega}_1(Y)$).

(2) $H_G^{r+1}(K,L;\tilde{\omega}_r(Y)) = 0$ for all $r < n$.

(3) $H_G^{r}(K,L;\tilde{\omega}_r(Y)) = 0$ for all $r < n$.

(4) $H_G^{r-1}(K,L;\tilde{\omega}_r(Y)) = 0$ for all $r < n$.

Numbers appearing in each statement indicate which of these assumptions are used. The results in this section are all easy applications of §1 to the study of extensions of equivariant maps and homotopies. The proofs will be omitted since they offer no difficulties.

Suppose first that we are given an equivariant map $f: L \to Y$.

(2.1) Lemma. (1,2) There exists an equivariant extension f_n of f to $K^n \cup L$.

(2.2) Lemma. (1,3) If f_n and g_n are equivariant extensions of f to $K^n \cup L$ then $[c_{f_n}] = [c_{g_n}]$.

(Hint: Use (2.1) to find a homotopy $f_{n-1} \sim g_{n-1}$ relative to L.)

(2.3) Definition. (1,2,3) Let $\gamma^{n+1}(f) \in H_G^{n+1}(K,L;\tilde{\omega}_n(Y))$ be the (unique) cohomology class $[c_{f_n}]$ for any equivariant extension f_n of f to $K^n \cup L$. $\gamma^{n+1}(f)$ is called the primary obstruction to extending f and is an invariant of the equivariant homotopy class of f.

(2.4) <u>Proposition.</u> <u>If</u> $k: K' \to K$ <u>is cellular and equivariant then</u> $\gamma^{n+1}(f \cdot k) = k^*(\gamma^{n+1}(f))$ <u>when this is defined.</u>

(This is also true without cellularity but we have not yet defined k^* in the general case.)

(2.5) <u>Theorem</u> (<u>Extension</u>). (1,2,3) <u>If we also have that</u> $H_G^{r+1}(K,L;\tilde{\omega}_r(Y)) = 0$ <u>for</u> $n < r < \dim(K-L)$ <u>then an equivariant map</u> $f: L \to Y$ <u>has an equivariant extension to</u> K <u>iff</u> $\gamma^{n+1}(f) = 0$.

Now suppose that we are given two equivariant maps $f,g: K \to Y$ such that $f|L = g|L$. These induce an equivariant map $f\#g: Q \to Y$ where $Q = (K\times\partial I) \cup (L\times I)$.

There is a natural isomorphism

(2.6) $\lambda: H_G^n(K,L;\tilde{\omega}_n(Y)) \xrightarrow{\approx} H_G^{n+1}(K\times I,Q;\tilde{\omega}_n(Y))$

(induced by the obvious isomorphism on the cochain level). We define, under conditions (1,3,4):

(2.7) $$\omega^n(f,g) = \lambda^{-1}(\gamma^{n+1}(f\#g))$$

and note that

(2.8) $$\omega^n(f,g) + \omega^n(g,h) = \omega^n(f,h)$$

and

(2.9) $$\omega^n(f\circ k,g\circ k) = k^*(\omega^n(f,g))$$

(where $k: (K',L') \to (K,L)$ is cellular and equivariant) when this is defined.

An application of (2.5) to this situation yields:

(2.10) Theorem (Homotopy). (1,3,4) If we also have that $H_G^r(K,L;\tilde{\omega}_r(Y)) = 0$ for $n < r \leq \dim(K-L)$ and if $f,g\colon K \to Y$ are equivariant with $f|L = g|L$, then f and g are equivariantly homotopic (relative to L) iff $\omega^n(f,g) = 0$.

A standard argument now proves the following result:

(2.11) Theorem (Classification). Assume that (1) holds and also that

$$\begin{cases} H_G^r(K,L;\tilde{\omega}_r(Y)) = 0 = H_G^{r-1}(K,L;\tilde{\omega}_r(Y)) \text{ for } r < n \\ H_G^r(K,L;\tilde{\omega}_r(Y)) = 0 = H_G^{r+1}(K,L;\tilde{\omega}_r(Y)) \text{ for } r > n. \end{cases}$$

Let $f\colon K \to Y$ be an equivariant map. Then the equivariant homotopy classes (relative to L) of maps $g\colon K \to Y$ (with $g|L = f|L$) are in one-one correspondence with the elements of

$$H_G^n(K,L;\tilde{\omega}_n(Y))$$

and $g \to \omega^n(g,f)$ is such a correspondence.

As a matter of notation, we shall use double brackets: $[[X;Y]]$, where X and Y are G-spaces, to denote the equivariant homotopy classes of (equivariant) maps $X \to Y$. Thus, for $L = \emptyset$, the conclusion of (2.11) states that $[[g]] \leftrightarrow \omega^n(g,f)$ is a one-one correspondence

$$[[K;Y]] \approx H_G^n(K;\tilde{\omega}_n(Y)).$$

3. The characteristic class of a map

In this section we assume that Y is a G-space with base point $y_0 \in Y^G$ such that

$$\tilde{\omega}_q(Y,y_0) = 0 \text{ for } q < n,$$

for a given integer $n \geq 1$. If $n = 1$, we assume that $\tilde{\omega}_1(Y,y_0)$ (that is, each $\pi_1(Y^H,y_0)$) is *abelian*.

Let K be a G-complex and let 0 denote the constant (equivariant) map $0\colon K \to y_0 \in Y$. For any equivariant map $f\colon K \to Y$ we define the characteristic class of f to be

$$(3.1) \qquad \chi^n(f) = \omega^n(f,0) \in H^n_G(K;\tilde{\omega}_n(Y)).$$

If $k\colon K' \to K$ is cellular and equivariant then by (2.9)

$$\chi^n(f\circ k) = k^*(\chi^n(f)).$$

(The cellularity condition is unnecessary as will follow from later facts.)

The following four results are standard and immediate consequences of the definitions and of §2. We shall omit their proofs:

(3.2) *Proposition. If* $H^r_G(K;\tilde{\omega}_r(Y)) = 0$ for $r > n$ *then two maps* $f,g\colon K \to Y$ *are homotopic iff* $\chi^n(f) = \chi^n(g)$.

(3.3) *Theorem. If* (K,L) *is a* G-*complex pair and* $f\colon L \to Y$ *is given with characteristic class* $\chi^n(f) \in H^n_G(L;\tilde{\omega}_n(Y))$, *then the primary obstruction to extending* f *to* K *equivariantly is*

$$\gamma^{n+1}(f) = \delta^*(\chi^n(f))$$

where $\delta^*\colon H^n_G(L;\tilde{\omega}_n(Y)) \to H^{n+1}_G(K,L;\tilde{\omega}_n(Y))$ *is the coboundary*.

(3.4) Corollary. *If* $H_G^{r+1}(K,L;\tilde{\omega}_r(Y)) = 0$ *for* $r > n$ *then an equivariant map* $f\colon L \to Y$ *has an equivariant extension to* K *iff*

$$\chi^n(f) \in \operatorname{Im}[i^*\colon H_G^n(K;\tilde{\omega}_n(Y)) \to H_G^n(L;\tilde{\omega}_n(Y))].$$

(3.5) Theorem. *If* $f,g\colon K \to Y$ *are equivariant and if* $f|L = g|L$, *then*

$$\chi^n(f) - \chi^n(g) = j^*(\omega^n(f,g)).$$

(Here $\chi^n(f)$ and $\chi^n(g)$ are in $H_G^n(K;\tilde{\omega}_n(Y))$, $\omega^n(f,g)$ is in $H_G^n(K,L;\tilde{\omega}_n(Y))$ and j^* is induced by $(K,\phi) \to (K,L)$.)

We conclude this section with some remarks on the case in which Y is, itself, a G-*complex*. These remarks will not be used in any essential way elsewhere in these notes. The identity $1\colon Y \to Y$ yields a class

$$\chi^n(Y) = \chi^n(1) = \omega^n(1,0) \in H_G^n(Y;\tilde{\omega}_n(Y)),$$

which is the primary obstruction to equivariantly contracting Y and is called the characteristic class of Y.

For any $f\colon K \to Y$ we obviously have

$$\chi^n(f) = f^*(\chi^n(Y)). \qquad (3.6)$$

By Chap. I, (10.5) we have that

$$H_G^n(Y;\tilde{\omega}_n(Y)) \approx \operatorname{Hom}(\tilde{\omega}_n(Y),\tilde{\omega}_n(Y))$$

and it can be shown that under this isomorphism $\chi^n(Y)$ corresponds to the identity homomorphism. (Perhaps the easiest way to prove this is to note that Y has the equivariant homotopy type of a G-complex which has no cells in dimensions between 0 and n, and then to prove the result in this case. See §7.) This is, of course, an important result since it allows the computation of the characteristic class.

4. Hopf G-spaces

Let Y be a G-space with base point y_0. Let G act diagonally on $Y \times Y$, that is, $g(y,y') = (gy,gy')$. Such a space Y together with a base point preserving equivariant map $\theta\colon Y\times Y \to Y$ is said to be a Hopf G-space if the restriction $Y \vee Y \to Y$ of θ is equivariantly homotopic to $1 \vee 1$. This obviously implies that Y^H, for $H \subset G$, is a Hopf-space.

For example, if Y is any G-space with base point, then the loop space ΩY is a Hopf G-space, where the action of G on a loop, or generally on a path, $f\colon I \to Y$, is defined by $g(f)(t) = g(f(t))$.

Let us denote the product $\theta(y,y')$ by $y \,\square\, y'$ in a given Hopf G-space Y. Let (K,L) be a pair of G-complexes and let $\varphi, \psi\colon K^n \cup L \to Y$ be equivariant, where Y is (also) as in §1. We have the map

$$\varphi \,\square\, \psi\colon K^n \cup L \to Y$$

defined by $(\varphi \,\square\, \psi)(x) = \varphi(x) \,\square\, \psi(x)$. Since addition in the homotopy groups of a Hopf-space is induced by the Hopf-space operation, as is well-known, and since each Y^{G_σ} is a Hopf-space, it follows immediately that

$$c_{\varphi\square\psi} = c_\varphi + c_\psi$$

in $C_G^{n+1}(K,L;\tilde{\omega}_n(Y))$.

It follows immediately that in the situation of (3.1), with Y a Hopf G-space and $f,f'\colon K \to Y$ equivariant, we have

$$(4.1) \qquad \chi^n(f \,\square\, f') = \chi^n(f) + \chi^n(f').$$

5.* Equivariant deformations and homotopy type

In this section we shall prove some elementary facts concerning equivariant deformation. These results could be encompassed in an obstruction theory of deformation (which contains the obstruction theory of extensions) but we have chosen not to do so.

Let $Y \supset B$ be a pair of G-spaces and assume that

(5.1) $\tilde{\omega}_q(Y,B) = 0$ for all $0 \leq q \leq n$,

in the sense that, for every subgroup $H \subset G$, every map $(B^q, S^{q-1}) \to (Y^H, B^H)$ is deformable, through such maps, to a map into B^H. (We allow the case $n = \infty$.)

(5.2) Lemma. Let (K,L) be a pair of G-complexes with $\dim(K-L) \leq n$ and let $\varphi: K,L \to Y,B$ be an equivariant map. Then φ is equivariantly homotopic relative to L to a map into B.

Proof. Consider $K \times I$. We wish to extend the map $\varphi | L \times I \cup \varphi \times \{0\}$ on $L \times I \cup K \times \{0\}$ to $K \times I$ such that $K \times \{1\}$ goes into B. The extension is defined inductively on the $K^n \times I$ and proceeds much as in the proof of (1.2). The details are omitted.

As above, double brackets $[[X;Y]]$ denote the set of equivariant homotopy classes of equivariant maps $X \to Y$, where X and Y are G-spaces.

(5.3) Corollary. Inclusion $i: B \to Y$ induces a one-one correspondence

$$i_{\#}: [[K;B]] \xrightarrow{\approx} [[K;Y]]$$

for every G-complex K with $\dim K < n$.

Proof. $i_\#$ is onto by (5.2). If $f: K \to B$ can be equivariantly deformed, through Y, to $g: K \to B$, then by (5.2) the homotopy may be deformed, relative to the ends, into B. This shows that $i_\#$ is one-one.

(5.4) Theorem. Let $f: Y \to Y'$ be an equivariant map of G-spaces such that $f_\#: \tilde{\omega}_q(Y) \approx \tilde{\omega}_q(Y')$ for all $q \geq 0$. Then

$$f_\#: [[K;Y]] \to [[K;Y']]$$

is a one-one correspondence for every G-complex K.

Proof. Let M_f be the mapping cylinder of f, with the natural G-action. M_f and Y' have the same equivariant homotopy type so that f may be replaced by the inclusion $i: Y \to M_f$. The hypothesis implies easily that $\tilde{\omega}_q(M_f, Y) = 0$ for all $q \geq 0$. Thus the result follows from (5.3).

(5.5) Corollary. If $\varphi: K \to K'$ is an equivariant map between two G-complexes such that $\varphi_\#: \tilde{\omega}_q(K) \approx \tilde{\omega}_q(K')$ for all $q \geq 0$ then φ is an equivariant homotopy equivalence.

Proof. $\varphi_\#: [[K';K]] \xrightarrow{\approx} [[K';K']]$ by (5.4). Let $\psi: K' \to K$ represent $\varphi_\#^{-1}(1)$. That is, $\varphi\psi: K' \to K'$ is equivariantly homotopic to the identity. Clearly $\psi_\# = \varphi_\#^{-1}$ is bijective so that there is (similarly) a $\theta: K \to K'$ with $\psi\theta \sim 1$ (equivariantly). Then $\theta \sim \varphi\psi\theta \sim \varphi$ so that $\psi\varphi \sim \psi\theta \sim 1$ as was to be shown.

(5.6) Proposition. Every equivariant map $f: K_1 \to K_2$ between two G-complexes is equivariantly homotopic to a cellular map. An equivariant homotopy between cellular maps may be

<u>deformed equivariantly, relative to the ends, into a cellular homotopy</u>.

<u>Proof</u>. This is an easy consequence of (5.2) using $(Y,B) = (K_2, K_2^n)$.

This result can be used to extend the definition of the induced cohomology homomorphism of an equivariant map $K_1 \to K_2$ to arbitrary (non-cellular) maps. Another method of doing this is given in §6.

6. Eilenberg-MacLane G-complexes

Let $\tilde{\omega}$ be any element of the abelian category $\mathcal{C}_G$. A G-space of <u>type</u> $(\tilde{\omega},n)$ is defined to be a G-space Y with

$$\tilde{\omega}_q(Y,y_0) = \begin{cases} 0 & \text{for } q \neq n \\ \tilde{\omega} & \text{for } q = n, \end{cases}$$

where $y_0 \in Y^G \neq \emptyset$.

For such a space (2.11) provides a one-one correspondence

(6.1) $$[[K;Y]] \approx H_G^n(K;\tilde{\omega})$$

for all G-complexes K, given by

$$[[f]] \leftrightarrow \omega^n(f,0) = \chi^n(f)$$

(where 0 denotes the constant map $K \to y_0$ and the notation on the right is from §3). Moreover, if $\varphi\colon K \to K'$ is cellular and equivariant then, by (2.9),

(6.2) $$\begin{array}{ccc} [[K';Y]] & \xrightarrow{\approx} & H_G^n(K';\tilde{\omega}) \\ \downarrow \varphi^{\#} & & \downarrow \varphi^{*} \\ [[K;Y]] & \xrightarrow{\approx} & H_G^n(K;\tilde{\omega}) \end{array}$$

commutes, where $\varphi^{\#}([[f]]) = [[f\circ\varphi]]$. Thus (6.1) is a natural equivalence of functors.

Note that if Y is a G-space of type $(\tilde{\omega},n)$ then the loop space ΩY (see §4) has type $(\tilde{\omega},n-1)$. This is an immediate consequence of the obvious fact that $(\Omega Y)^H = \Omega(Y^H)$.

If Y is a Hopf G-space then we can define an addition in $[[K,Y]]$ by

$$(6.3) \qquad [[f]] + [[f']] = [[f \,\square\, f']].$$

Then, by (4.1), the correspondence (6.1) preserves addition. Thus, in this case, if we are given an equivariant map $\varphi: K \to K'$ (not necessarily cellular) we can define $\varphi^*: H^n_G(K';\tilde{\omega}) \to H^n_G(K;\tilde{\omega})$ by commutativity of (6.2), since $\varphi^{\#}$ is always defined. The obvious additivity of $\varphi^{\#}$ implies that φ^* is a homomorphism. Thus, in this way, we can dispense with the definitions of φ^* in Chap. I, §7 as well as Proposition (5.6) (used to extend the definition of φ^* to non-cellular maps).

We shall now show how to construct a G-complex K of type $(\tilde{\omega},n)$ for any $\tilde{\omega} \in \mathcal{C}_G$ and $n \geq 1$. We shall restrict our attention, for convenience only, to the case $n > 1$. This is not much loss of generality since ΩK has type $(\tilde{\omega},1)$ when K has type $(\tilde{\omega},2)$. The construction is based on the following two lemmas which use the notation of Chap. I, §9, 10.

First we shall introduce some further notation. If T is a G-set, T^+ is T together with a disjoint base point, S^qT^+ is the qth reduced suspension of T^+ (that is, the one point union of q-spheres, one for each member of T), and CS^qT^+ is the reduced cone of this (that is, the one point union of (q+1)-cells, one for each member of T). Note that there are natural isomorphisms (for $q > 1$)

(6.4) $$F_T \approx \tilde{\omega}_q(S^qT^+) \approx \underline{C}_q(S^qT^+;Z) \approx \underline{H}_q(S^qT^+;Z)$$

of elements of $\mathcal{C}_G$.

(6.5) Lemma. *Let* $q > 1$ *and let* Y *be a* G-*space with base point* y_0 *and with* $\tilde{\omega}_0(Y,y_0) = 0 = \tilde{\omega}_1(Y,y_0)$. *Then for any* G-*set* T, *the assignment to an equivariant homotopy class* [[f]] (*of a map* $f: S^qT^+ \to Y$) *of the induced morphism* $f_\#: F_T \approx \tilde{\omega}_q(S^qT^+) \to \tilde{\omega}_q(Y)$ *in* $\mathcal{C}_G$ *is a one-one correspondence*

$$[[S^qT^+,Y]] \xrightarrow{\approx} \text{Hom}\,(F_T,\tilde{\omega}_q(Y)).$$

In particular, every morphism $\alpha: F_T \to \tilde{\omega}_q(Y)$ *in* $\mathcal{C}_G$ *is represented by an equivariant map* $f: S^qT^+ \to Y$ *and* f *is equivariantly extendible to* $CS^qT^+ \to Y$ *iff* $\alpha = f_\#$ *is trivial*.

Proof. A direct proof of this should be fairly obvious. However, we note that it is, in fact, a special case of the equivariant homotopy classification theorem (2.11). That is, take $K = S^qT^+$, let L be the base point, and let $0: K \to Y$ be the constant map into y_0. The conditions of (2.11) are satisfied for $n = q$ since, in fact, K has no cells in dimensions other than 0 and q. The classification assigns to an equivariant map $f: K \to Y$ the class $\omega^q(f,0)$ in

$$H^q_G(S^qT^+;\tilde{\omega}_q(Y)) \approx \text{Hom}(\underline{H}_q(S^qT^+;Z),\tilde{\omega}_q(Y))$$

$$\approx \text{Hom}(\tilde{\omega}_q(S^qT^+),\tilde{\omega}_q(Y))$$

(see (9.5) of Chap. I). It is obvious from the definition of $\omega^n(f,0)$ that the corresponding homomorphism $\tilde{\omega}_q(S^qT^+) \to \tilde{\omega}_q(Y)$ is precisely the induced map $f_\#$.

(6.6) Lemma. Let $q > 1$ and let Y be a G-space with base point y_0 and with $\tilde{\omega}_0(Y,y_0) = 0 = \tilde{\omega}_1(Y,y_0)$. Let $f\colon S^qT^+ \to Y$ be an equivariant base point preserving map and let $Y' = Y \cup_f CS^qT^+$ be the (reduced) mapping cone of f with the obvious G action. Let $i\colon Y \to Y'$ denote the inclusion. Then we have the following facts:

(1) $i_\#\colon \tilde{\omega}_r(Y) \to \tilde{\omega}_r(Y')$ is an isomorphism for $r < q$.

(2) $i_\#\colon\tilde{\omega}_q(Y) \to \tilde{\omega}_q(Y')$ is an epimorphism with Kernel $i_\# = \operatorname{Im}\{f_\#\colon F_T \approx \tilde{\omega}_q(S^qT^+) \to \tilde{\omega}_q(Y)\}$.

Proof. For $H \subset G$ it is clear that Y'^H is just the mapping cone of $(S^qT^+)^H \to Y^H$. But $(S^qT^+)^H = S^q(T^H)^+$. Thus $i_\#(G/H)\colon \pi_r(Y^H) \to \pi_r(Y'^H)$ is induced by the inclusion of Y^H in the mapping cone of the restriction of $f\colon S^q(T^H)^+ \to Y^H$. Similarly $f_\#(G/H)\colon \pi_r((S^qT^+)^H) \to \pi_r(Y^H)$ is induced by the restriction of f. Since (1) and (2) are true iff the corresponding statements for the values on each $G/H \in \mathcal{C}_G$ are true, and since these corresponding statements are known results (see Hu, Homotopy Theory, p. 168) concerning (non-equivariant) attaching of cells, the lemma follows.

Using these two lemmas, the construction of $K(\tilde{\omega},n)$ complexes is now quite straightforward. Thus let T and R be G-sets such that there is an exact sequence

$$F_R \xrightarrow{\alpha} F_T \xrightarrow{\beta} \tilde{\omega} \to 0$$

in $\mathcal{C}_G$ (see Chap. I, §10). Let $n > 1$ and put $K^n = S^nT^+$. Let

$$f\colon S^nR^+ \to S^nT^+$$

be an equivariant map inducing α (via $F_T \approx \tilde{\omega}_n(S^nT^+)$, etc.). This exists by (6.5). Let $K^{n+1} = K^n \cup_f CS^nR^+$. By (6.6) we have

$$\begin{cases} \tilde{\omega}_n(K^{n+1}) \approx \tilde{\omega} \\ \tilde{\omega}_r(K^{n+1}) = 0 \text{ for } r < n. \end{cases}$$

If K^q has been constructed to be a G-complex of dimension q ($q \geq n+1$) such that

$$(6.7) \qquad \begin{cases} \tilde{\omega}_n(K^q) \approx \tilde{\omega} \\ \tilde{\omega}_r(K^q) = 0 \text{ for } r < n \text{ and } n < r < q \end{cases}$$

let V be a G-set such that there is an epimorphism

$$F_V \xrightarrow{\gamma} \tilde{\omega}_q(K^q).$$

Let $v\colon S^qV^+ \to K^q$ be an equivariant map inducing γ and let $K^{q+1} = K^q \cup_v S^qV^+$. Then, by (6.6), K^{q+1} satisfies (6.7) with q replaced by $q+1$. Let $K = \bigcup_q K^q$. This is clearly a G-complex of type $(\tilde{\omega},n)$.

7.* n-connected G-complexes

The method of killing the groups $\tilde{\omega}_q$ used in the construction of $K(\tilde{\omega},n)$ in the last section is, of course, an important tool. We shall use it here in a rather straightforward way to prove the following result:

(7.1) *Proposition. Let K be a G-complex with $\tilde{\omega}_q(K) = 0$ for all $0 \leq q < n$. Then K has the same equivariant homotopy type as a G-complex with no cells in dimensions q for $0 < q < n$.*

Proof. Let $L = K^{n-1}$. Then the inclusion $L \to K$ is equivariantly homotopic to a constant map, by (2.10). That is, K is an equivariant retract of $K \cup C_L$. But $K \cup C_L$ has the same equivariant homotopy type as K/L. Thus there exist equivariant maps

$$K \xrightarrow{\varphi} K/L \xrightarrow{\psi} K$$

with $\psi\varphi$ equivariantly homotopic to 1. Clearly K/L has no q-cells for $0 < q < n$ so that $\tilde{\omega}_q(K/L) = 0$ for $q < n$.

Suppose that for some $q \geq n$ we have constructed a G-complex $K_q \supset K/L$ and an equivariant map $\psi_q: K_q \to K$ with $\psi_q\varphi = \psi\varphi$ such that $(\psi_q)_\#: \tilde{\omega}_r(K_q) \to \tilde{\omega}_r(K)$ is a monomorphism for $r < q$. Let T be a G-set and

$$\alpha: F_T \to \operatorname{Ker}\{(\psi_q)_\#: \tilde{\omega}_q(K_q) \to \tilde{\omega}_q(K)\}$$

an epimorphism in $\mathcal{C}_G$. Let

$$f: S^qT^+ \to K_q$$

be an equivariant map inducing α. f may be assumed to be cellular by (5.6) (or merely because $\tilde{\omega}_q(K_q^q) \to \tilde{\omega}_q(K_q)$ is an epimorphism). Let $K_{q+1} = K_q \cup_f CS^qT^+$. By (6.5), ψ_q extends to $\psi_{q+1}: K_{q+1} \to K$ and, by (6.6), $(\psi_{q+1})_\#: \tilde{\omega}_r(K_{q+1}) \to \tilde{\omega}_r(K)$ is a monomorphism for $r \leq q$. Let (K', ψ') be the union of the (K_q, ψ_q).

Thus we obtain a G-complex $K' \supset K/L$ with no q-cells for $0 < q < n$ and equivariant maps

$$K \xrightarrow{\varphi} K' \xrightarrow{\psi'} K$$

with $\psi'\varphi = \psi\varphi \sim 1$. Also

$$\psi'_\#: \tilde{\omega}_*(K') \to \tilde{\omega}_*(K),$$

being a monomorphism with $\psi'_\# \varphi_\# = 1$, must be an isomorphism and it follows from (5.5) that K and K' have the same equivariant homotopy type.

Chapter III. Function Spaces, Fibrations and Spectra

In this chapter we shall gather some miscellaneous items. The first and third sections contain some definitions and terminology that will be used later.

1. Function spaces

In this section we work in the category of G-spaces with base point. The group G is arbitrary and need not be finite.

If X and Y are G-spaces we let

$$F(X,Y)$$

denote the space of all (base point preserving) maps from X to Y in the compact-open topology. F(X,Y) is a G-space with the following G-action: If $f\colon X \to Y$ and $g \in G$ we put

$$g(f)(x) = g(f(g^{-1}x)).$$

The set $F(X,Y)^G$ of stationary points of G on F(X,Y) is just the set of equivariant maps from X to Y. Thus we put

(1.1) $$E(X,Y) = F(X,Y)^G.$$

Note that the reduced join

$$X \wedge Y = X \times Y / X \vee Y$$

of G-spaces has a natural G-action induced from the diagonal action on X×Y. Also recall that, for Y locally compact, there is a homeomorphism

(1.2) $$F(X \wedge Y, Z) \xrightarrow{\approx} F(X, F(Y,Z))$$

taking f into $\bar{f}$ defined by $(\bar{f}(x))(y) = f(x \wedge y)$. Note that

$$(g(\overline{f})(x))(y) = (g(\overline{f}(g^{-1}x)))(y)$$
$$= g[(\overline{f}(g^{-1}x))(g^{-1}y)] = g[f(g^{-1}x \wedge g^{-1}y)]$$
$$= g[f(g^{-1}(x \wedge y))] = g(f)(x \wedge y) = (\overline{g(f)}(x))(y),$$

that is, $g(\overline{f}) = \overline{g(f)}$, which means that (1.2) is equivariant.

In particular (1.2) induces a homeomorphism

(1.3) $$E(X \wedge Y, Z) \xrightarrow{\approx} E(X, F(Y,Z)),$$

when Y is locally compact.

If G acts trivially on X, so that $X = X^G$, then clearly $E(X,Y) = F(X,Y^G)$. In particular,

(1.4) $$E(X,F(Y,Z)) \approx F(X,E(Y,Z)) \text{ when } X = X^G.$$

Now the reduced suspension $SX = S \wedge X$ is a G-space, the action on the factor $S = S^1$ being trivial. Similarly, the loop space $\Omega X = F(S,X)$ is a G-space, as above. Thus (1.2) provides the equivariant homeomorphism

(1.5) $$F(SX,Y) \approx F(X,\Omega Y).$$

The comultiplication $SX \to SX \vee SX$ and the loop multiplication on ΩY induce Hopf G-space structures (see Chap. II, §4) on $F(SX,Y)$ and $F(X,\Omega Y)$ and it is well-known, and elementary, that these structures correspond under (1.5). In particular, passing to sets of stationary points, we have the isomorphism

(1.6) $$E(SX,Y) \approx E(X,\Omega Y)$$

of Hopf-spaces.

It is easy to see that (1.6) preserves equivariant homotopies. Thus, denoting equivariant homotopy classes by double square brackets, as before, we have the one-one correspondence

$$(1.7) \qquad [[SX;Y]] \leftrightarrow [[X;\Omega Y]]$$

which preserves addition.

F(SX,ΩY) possesses two Hopf G-space structures. Let us denote the one induced by comultiplication in SX by $\circ$ and that induced by loop multiplication by $\square$. Then it is well-known, and easily checked, that we have the identity

$$(f\circ g)\square(h\circ k) = (f\square h)\circ(g\square k).$$

This identity is, of course, also satisfied on the fixed point set E(SX,ΩY), and also for the induced multiplications on [[SX;ΩY]]. But the latter set has an identity e for both $\circ$ and $\square$ and we have

$$\alpha\square\beta = (e\circ\alpha)\square(\beta\circ e) = (e\square\beta)\circ(\alpha\square e) = \beta\circ\alpha$$

and $\alpha\square\beta = (\alpha\circ e)\square(e\circ\beta) = (\alpha\square e)\circ(e\square\beta) = \alpha\circ\beta$ so that

$$(1.8) \qquad \alpha\circ\beta = \beta\circ\alpha = \alpha\square\beta = \beta\square\alpha$$

on [[SX;ΩY]]. (The statement on E(SX,ΩY) is that the corresponding maps are homotopic.)

It should be noted that when X is <u>locally compact</u>, we can improve these remarks as follows. We have, by (1.3) and (1.4),

$$E(SX,Y) \approx E(S,F(X,Y)) = F(S,E(X,Y)) = \Omega E(X,Y),$$

Also $[[X;Y]] = \pi_0 E(X,Y)$ so that we obtain

$$(1.9) \qquad [[X;\Omega Y]] \approx [[SX;Y]] \approx \pi_1(E(X,Y)).$$

Similarly,

$$(1.10) \qquad [[X;\Omega^n Y]] \approx [[S^n X;Y]] \approx \pi_n(E(X,Y)).$$

2.* The Puppe sequence

In this section we consider only spaces with base points. Let $f: X \to Y$ be an equivariant map between two G-spaces. Let $C_f = CX \cup_f Y$ be the reduced mapping cone of f with the obvious G-action, and let $j: Y \to C_f$ be the canonical inclusion. It is clear that, for any G-space Z, the sequence

(2.1) $$[[C_f;Z]] \xrightarrow{j^\#} [[Y;Z]] \xrightarrow{f^\#} [[X;Z]]$$

of sets with base points is exact. It can be shown that the mapping cone C_j of j has the same homotopy type as does SX (see Puppe, Math. Zeitschrift, 69 (1958) pp. 299-344). The proof of this is sufficiently canonical to be equivariant and we shall not give the details of this here. Thus C_j has the equivariant homotopy type of SX.

As in [Puppe, loc. cit.] we combine (2.1) with the similar sequence for $Y \xrightarrow{j} C_f \to C_j \sim SX$ and continue this process to finally obtain a long exact sequence

(2.2) $$\dots \to [[S^n C_f;Z]] \to [[S^n Y;Z]] \to [[S^n X;Z]] \to [[S^{n-1} C_f;Z]] \to \dots$$

3. G-spectra

In this section we work with the category of spaces (or G-spaces) with base points. By a G-spectrum we mean a collection $\underline{Y} = \{Y_n | n \in Z\}$ of G-spaces, together with equivariant maps

(3.1) $$\varepsilon_n: SY_n \to Y_{n+1}$$

or, by (1.6), of equivariant maps $Y_n \to \Omega Y_{n+1}$. We note that it is sufficient to have Y_n defined for $n \geq n_0$ and let Y_n be a point for $n < n_0$.

If $\underline{\underline{Y}}$ is a G-spectrum and if X is a locally compact G-space, then

$$\underline{\underline{F}}(X,\underline{\underline{Y}})$$

denotes the G-spectrum consisting of the G-spaces $F(X,Y_n)$ and the equivariant maps defined by the composition

$$F(X,Y_n) \to F(X,\Omega Y_{n+1}) \xrightarrow{\approx} F(SX,Y_{n+1}) \xrightarrow{\approx} \Omega F(X,Y_{n+1}).$$

In particular, $\Omega\underline{\underline{Y}}$ is a G-spectrum.

Note that if $\underline{\underline{Y}}$ is a G-spectrum then $\underline{\underline{Y}}^G = \{Y_n^G\}$ is a spectrum. In particular, for X locally compact,

$$\underline{\underline{E}}(X,\underline{\underline{Y}}) = \underline{\underline{F}}(X,\underline{\underline{Y}})^G$$

is a spectrum consisting of the spaces $E(X,Y_n)$.

For a detailed treatment of spectra see G. Whitehead, Generalized homology theories, Trans. A.M.S. 102 (1962), pp. 227-283.

We shall list below some examples of G-spectra:

(1) If Y is a G-space (with base point) and n is an integer, let $Y_n = Y$ and $Y_{n+k} = S^kY$ with the obvious maps $SY_m \to Y_{m+1}$. This forms a G-spectrum $\underline{\underline{S}}(Y,n)$.

(2) If $\rho\colon G \to O(r)$ is a representation of G on $\mathbb{R}^r$ then $\rho \oplus 1$ defines an action (with base point) on S^r and thus defines a G-space S^r_ρ. We denote the G-spectrum $\underline{\underline{S}}(S^r_\rho,r)$ by $\underline{\underline{S}}(\rho)$.

(3) Let $G = Z_2$ and let ρ be the representation defined by the antipodal map in $\mathbb{R}^r$. We denote the G-spectrum $\underline{\underline{S}}(\rho)$ by $\underline{\underline{S}}(r)$. Here the n-th G-space in $\underline{\underline{S}}(r)$, for $n \geq r$, is S^n with a standard involution which leaves S^{n-r} stationary. Thus $\underline{\underline{S}}(r)$ may be called the spectrum of spheres with stationary points of codimension r.

(4) Let $\tilde{\omega} \in \mathcal{C}_G$ and let Y_n be a G-complex of type $(\tilde{\omega},n)$. Since ΩY_{n+1} has type $(\tilde{\omega},n)$ there is a map $\eta_n: Y_n \to \Omega Y_{n+1}$ whose characteristic class

$$\chi^n(\eta_n) \in H^n_G(Y_n, \tilde{\omega}_n(\Omega Y_{n+1})) \approx \mathrm{Hom}(\tilde{\omega}_n(Y_n), \tilde{\omega}_n(\Omega Y_{n+1}))$$

corresponds to the identity $1: \tilde{\omega} \to \tilde{\omega}$ (via given isomorphisms $\omega_n(Y_n) \approx \tilde{\omega}$ and $\tilde{\omega}_n(\Omega Y_{n+1}) \approx \tilde{\omega}$). Thus we obtain a spectrum denoted by $K(\tilde{\omega})$, the Eilenberg-MacLane G-spectrum of $\tilde{\omega}$.

4.* G-fiber spaces

Let $\pi: X \to Y$ be an equivariant map between two G-spaces, where G is finite. We shall say that π is a G-fiber map if it has the equivariant homotopy lifting property with respect to G-complexes. That is, if K is a G-complex, $f: K \to X$ is equivariant and $F: K \times I \to Y$ is equivariant with $F(k,0) = \pi f(k)$, then there exists an equivariant map

$$F': K \times I \to X \quad \text{with} \quad F = \pi F' \quad \text{and} \quad F'(k,0) = f(k).$$

(4.1) <u>Theorem.</u> $\pi: X \to Y$ <u>is a</u> G-<u>fiber map iff</u> $\pi|X^H: X^H \to Y^H$ <u>is a (Serre) fibration for every</u> $H \subset G$.

<u>Proof.</u> If K is any complex then any map $K \to X^H$ has a unique equivariant extension to $f: K\times(G/H) \to X$ (where the action of G on K is trivial). Moreover, an equivariant map $K\times(G/H) \to X$ must take $K\times(H/H)$ into X^H. It follows easily that $X^H \to Y^H$ must be a fibration when $X \to Y$ is a G-fibration.

Suppose that each $X^H \to Y^H$ is a fibration. Let K be a G-complex, $f: K \to X$ equivariant and $F: K\times I \to Y$ equivariant with $F(k,0) = \pi f(k)$ for each $k \in K$. We must construct $F': K\times I \to X$ equivariant with $F = \pi F'$ and $F'(k,0) = f(k)$. This will be done by induction on the skeletons of K. Suppose F' is defined on $K^{n-1}\times I$ and let σ be an n-cell of K. Let $H = G_\sigma$. Now $f: K(\sigma) \to X^H$, $F: K(\sigma)\times I \to Y^H$ and $F': (K^{n-1} \cap K(\sigma))\times I \to X^H$. Since $X^H \to Y^H$ is a fibration we may extend F' to a map $K(\sigma)\times I \to X^H$ covering F. There is then a unique equivariant extension of F' to the cells $g\sigma\times I$ for $g \in G$. If this construction is repeated for each orbit of G on the set of n-cells of K, we obtain the required extension of F' to $K^n\times I \to X$.

As an example, let Y be a G-space such that each Y^H is arcwise connected and let $y_0 \in Y^G$ be a base point. Then the space PY of paths on Y with initial point y_0 is a G-space and the canonical projection $\pi: PY \to Y$ is equivariant. Clearly $(PY)^H = P(Y^H)$ and the restriction $P(Y^H) \to Y^H$ of π is just the path-loop fibration of Y^H. Thus π is a G-fibration.

Suppose now that $\pi: X \to Y$ is a G-fibration. Let $x_0 \in X^G$ be a base point and put $y_0 = \pi(x_0)$. The G-space

$F = \pi^{-1}(y_0)$ is called the fiber of this fibration. As in the non-equivariant theory, we have an exact sequence

$$(4.2) \qquad \ldots \to \tilde{\omega}_n(F,x_0) \xrightarrow{i_\#} \tilde{\omega}_n(X,x_0) \xrightarrow{\pi_\#} \tilde{\omega}_n(Y,y_0) \xrightarrow{\partial_\#} \tilde{\omega}_{n-1}(F,x_0) \xrightarrow{i_\#} \ldots$$

In fact, the exactness of this sequence follows from the exactness of the homotopy sequences of the fibrations $X^H \to Y^H$ with fiber F^H. Of course one must show that $i_\#$, $\pi_\#$ and $\partial_\#$, which are defined so that their values on G/H are the corresponding homomorphisms for the fibration $X^H \to Y^H$, are in fact morphisms in $\mathcal{C}_G$. This is left to the reader.

Chapter IV. Generalized Equivariant Cohomology

In this chapter we show how to construct generalized equivariant cohomology theories, using G-spectra. We then show how any generalized theory is connected by a spectral sequence to the "classical" theory of Chapter I.

1. Equivariant cohomology via G-spectra

We work with the category of spaces with base points in this section. Let Y be a G-spectrum. Then for any G-space X we have homomorphisms

$$\eta_k: [[S^{k-n}X;Y_k]] \xrightarrow{S} [[S^{k-n+1}X;SY_k]] \xrightarrow{\epsilon_{k\#}} [[S^{k-n+1}X;Y_{k+1}]].$$

Thus, with these maps, the groups $[[S^{k-n}X;Y_k]]$ form a direct system and we define

$$(1.1) \qquad \tilde{H}^n_G(X;\underline{Y}) = \lim_k [[S^{k-n}X;Y_k]] = \lim_k [S^kX;Y_{m+k}]].$$

Note that if X is locally compact then this is the same as

$$(1.2) \qquad \pi_{-n}(\underline{E}(X,\underline{Y})) = \lim_k \pi_{k-n}(E(X,Y_k)).$$

Note that $[[S^kX;Y_{n+k}]] \approx [[X;\Omega^kY_{n+k}]]$. If $A \subset X$ is invariant under G, then for any G-space W there is the exact sequence

$$[[X \cup C_A;W]] \to [[X;W]] \to [[A;W]]$$

of (2.1) in Chapter III. If (X,A) is a pair of G-complexes, then $X \cup C_A$ has the same equivariant homotopy type as does X/A. Thus, taking $W = \Omega^kY_{n+k}$, and passing to the limit over k, we obtain the exact sequence

$$(1.3) \qquad \tilde{H}^n_G(X/A;\underline{Y}) \to \tilde{H}^n_G(X;\underline{Y}) \to \tilde{H}^n_G(A;\underline{Y})$$

on the category $\mathcal{S}_0$ of G-complexes with base point.

Using the natural homeomorphism $S^{k-n}X \approx S^{k-(n+1)}SX$ we obtain a natural isomorphism $s_k: [[S^{k-n}X;Y_k]] \overset{\approx}{\to} [[S^{k-(n+1)}SX;Y_k]]$. These commute with the n_k and hence define a natural isomorphism

$$s^*: \tilde{H}_G^n(X;\underline{Y}) \to \tilde{H}_G^{n+1}(SX,\underline{Y}).$$

We have shown that $\tilde{H}_G^*(X;\underline{Y})$ defines an equivariant cohomology theory on $\mathcal{B}_0$.

2. Exact couples

In this section we provide some background from the theory of exact couples. Let

(2.1)

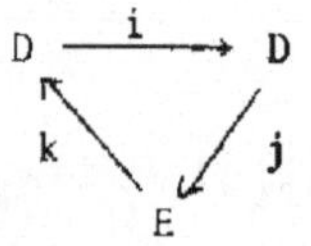

be an exact couple where E and D are bigraded, k has total degree 1 and i and j have total degree 0. Note that $(jk)^2 = 0$ and let H(E) be the homology of E with respect to the differential jk. The derived couple of (2.1) is

$$
\begin{array}{ccc}
iD & \overset{i'}{\longrightarrow} & iD \\
{}^{k'}\nwarrow & & \swarrow^{j'} \\
& H(E) &
\end{array}
$$

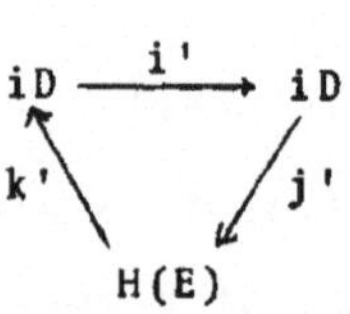

where $i' = i|iD$, j' is induced by ji^{-1} and k' is induced by k. Let $D_1 = D$ and $E_1 = E$. Iterating the above procedure we obtain the (r-1)st derived couple

$$\begin{array}{ccc} D_r & \xrightarrow{i_r} & D_r \\ & {\scriptstyle k_r}\nwarrow \quad \swarrow{\scriptstyle j_r} & \\ & E_r & \end{array}$$

where $E_r = H(E_{r-1})$ and $D_r = iD_{r-1} = i^{r-1}D$.

We shall now assume that

$$(2.2) \qquad \begin{cases} \deg i = (-1,1) \\ \deg j = (0,0) \\ \deg k = (1,0) \end{cases}$$

and it is then easy to check that

$$(2.3) \qquad \begin{cases} \deg i_r = (-1,1) \\ \deg j_r = (r-1,1-r) \\ \deg k_r = (1,0). \end{cases}$$

We let $d_r = j_r k_r$ which has degree $(r,1-r)$. The system $\{E_r^{p,q}\}$ together with the differentials d_r then form a spectral sequence.

We shall now assume that, for some integer N,

$$(2.4) \qquad \begin{cases} E^{p,q} = 0 & \text{for } p < 0 \text{ and for } p > N \\ D^{p,q} = 0 & \text{for } p < 0. \end{cases}$$

From the exact sequence

$$\cdots \to D^{p,q} \xrightarrow{j} E^{p,q} \xrightarrow{k} D^{p+1,q} \xrightarrow{i} D^{p,q+1} \xrightarrow{j} E^{p,q+1} \to \cdots$$

we see that

$$i\colon D^{p+1,q} \xrightarrow{\approx} D^{p,q+1} \qquad \text{for } p > N.$$

For $n = p+q$ we let J^n be a group which is isomorphic to $D^{p,q+1}$ for $p > N$ and let $\theta^{p,q+1}\colon J^n \to D^{p,q+1}$ be some isomorphism chosen so that

$$(2.5)\qquad \begin{array}{ccc} & \xrightarrow{\theta^{p+1,q}} & D^{p+1,q} \\ J^n & & \downarrow i \\ & \xrightarrow[\theta^{p,q+1}]{} & D^{p,q+1} \end{array}$$

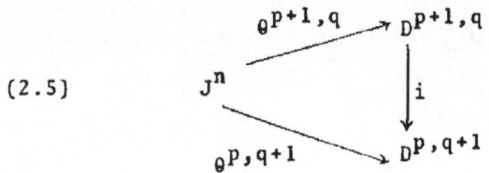

commutes. Following θ by iterates of i we have homomorphisms $\theta^{p,q+1}: J^n \to D^{p,q+1}$ defined for all p (with $n = p+q$) such that (2.5) commutes.

If $r > N$ we see that $d_r = 0$, since $E_r^{p,q} = 0$ for $p < 0$ and for $p > N$. Thus

$$E_r^{p,q} \approx E_{r+1}^{p,q} \approx \cdots$$

for $r > N$ and we let $E_\infty^{p,q}$ denote the common value. The (r-1)st derived couple has the form

$$\cdots i^{r-1}D^{p,q} \xrightarrow{j_r} E_r^{p,q} \xrightarrow{k_r} i^{r-1}D^{p+r,q-r+1} \xrightarrow{i} i^{r-1}D^{p+r+1,q-r} \to \cdots$$

Now $i^{r-1}D^{p,q} \subset D^{p-r+1,q+r-1} = 0$ for r sufficiently large and $i^{r-1}D^{p+r,q-r+1} = \operatorname{Im}\theta^{p+1,q} \subset D^{p+1,q}$ for r sufficiently large. Thus, for r large, this exact sequence has the form

$$(2.6)\qquad 0 \to E_\infty^{p,q} \to \operatorname{Im}\theta^{p+1,q} \xrightarrow{i} \operatorname{Im}\theta^{p,q+1} \to 0.$$

That is, we have an exact sequence

$$(2.7)\qquad 0 \to E_\infty^{p,q} \to \frac{J^{p+q}}{\ker\theta^{p+1,q}} \xrightarrow{i} \frac{J^{p+q}}{\ker\theta^{p,q+1}} \to 0\,.$$

Put

$$(2.8)\qquad J^{p,q} = \ker\{\theta^{p,q+1}: J^{p+q} \to D^{p,q+1}\}$$

so that (2.7) provides the isomorphism

$$(2.9)\qquad E_\infty^{p,q} \approx J^{p,q}/J^{p+1,q-1}\,.$$

Thus we have that the spectral sequence $E_r^{p,q}$ converges to the graded group associated with the (finite) filtration

$$\cdots \supset J^{p,q} \supset J^{p+1,q-1} \supset \cdots$$

of $J^{p+q} = D^{M+1,p+q-M}$ for $M \geq N$.

3. The spectral sequence of a filtered G-complex

Let K be a G-complex and let $\{K_r\}$ be a sequence of G-subcomplexes such that

$$(3.1) \qquad \begin{cases} K_r \subset K_{r+1} \\ K_{-1} = \emptyset \\ K_N = K \end{cases}$$

where N is some given integer.

Let $\{\mathcal{H}^*, \delta^*\}$ be any equivariant cohomology theory and put

$$(3.2) \qquad \begin{cases} E^{p,q} = \mathcal{H}^{p+q}(K_p, K_{p-1}) \\ D^{p,q} = \mathcal{H}^{p+q-1}(K_{p-1}). \end{cases}$$

Then the exact cohomology sequence of the pair (K_p, K_{p-1}) provides an exact couple

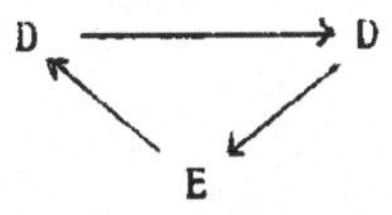

as in section 2.

The differential d_1 is the composition

$$(3.3) \qquad E_1^{p,q} = \mathcal{H}^{p+q}(K_p, K_{p-1}) \to \mathcal{H}^{p+q}(K_p) \xrightarrow{\delta} \mathcal{H}^{p+q+1}(K_{p+1}, K_p) = E_1^{p+1,q}.$$

And the spectral sequence converges to the graded group associated with the filtration

$$J^{p,q} = \ker\{\mathcal{H}^{p+q}(K) \to \mathcal{H}^{p+q}(K_{p-1})\}$$

of $J^{p+q} = \mathcal{H}^{p+q}(K)$.

4. The main spectral sequence

Let $\{\mathcal{H}^*, \delta^*\}$ be any equivariant cohomology theory and let K be a G-complex of dimension $N < \infty$. If K is not finite then we shall assume that $\mathcal{H}^*$ also satisfies the axiom:

(A) If S is a discrete G-set with orbits S_α then $\prod i_\alpha^*: \mathcal{H}^n(S) \to \prod \mathcal{H}^n(S_\alpha)$ is an isomorphism, where $i_\alpha: S_\alpha \to S$ is the inclusion.

Letting $K_p = K^p$, the p-skeleton of K, the preceding section provides a spectral sequence with

$$E_1^{p,q} = \mathcal{H}^{p+q}(K^p, K^{p-1}) \approx \mathcal{H}^{p+q}(K^p/K^{p-1}).$$

Now

$$K^p/K^{p-1} \approx S^p C_p^+$$

the p-th reduced suspension of the discrete G-set C_p^+ where C_p stands for the set of all p-cells of K. Thus

$$E_1^{p,q} \approx \tilde{\mathcal{H}}^{p+q}(S^p C_p^+) \approx \tilde{\mathcal{H}}^q(C_p^+) \approx \mathcal{H}^q(C_p).$$

Now let $h^q \in \mathcal{C}_G$ denote the coefficient system of Chapter I, section 4, example (1). That is

$$h^q(G/H) = \mathcal{H}^q(G/H) = \tilde{\mathcal{H}}^q((G/H)^+).$$

We shall define an isomorphism

$$\text{(4.1)} \qquad \alpha : \tilde{\mathcal{H}}^q(C_q^+) \xrightarrow{\approx} C_G^p(K; h^q)$$

as follows:

For $\sigma \in C_p$ let

$$i_\sigma : (G/G_\sigma)^+ \to C_p^+$$

be the equivariant map defined by $i_\sigma(gG_\sigma) = g\sigma \in C_p$. Also let

$$j_\sigma : C_p^+ \to (G/G_\sigma)^+$$

be defined by $j_\sigma(g\sigma) = gG_\sigma$ and $j_\sigma(\tau)$ = base point if τ is not in the orbit of σ. Note that

$$(4.2)\qquad \begin{cases} i_{g\sigma} = i_\sigma \hat{g} \\ j_{g\sigma} = \hat{g}^{-1} j_\sigma \\ j_\sigma i_\sigma = 1 \\ j_\tau i_\sigma = 0 \text{ (the base point) if } \tau \notin G(\sigma), \end{cases}$$

where $\hat{g} = R_g : G/G_{g\sigma} = G/gG_\sigma g^{-1} \to G/G_\sigma$. Also note that $i_\sigma j_\sigma$ is the identity on $G(\sigma)$ and collapses everything else to the base point.

We have the induced maps

$$\begin{cases} i_\sigma^* : \tilde{\mathcal{H}}^q(C_p^+) \to \tilde{\mathcal{H}}^q((G/G_\sigma)^+) = h^q(G/G_\sigma) \\ j_\sigma^* : \tilde{\mathcal{H}}^q((G/G_\sigma)^+) \to \tilde{\mathcal{H}}^q(C_p^+). \end{cases}$$

Define, for $\lambda \in \tilde{\mathcal{H}}^q(C_p^+)$ and $\sigma \in C_p$,

$$(4.3)\qquad \alpha(\lambda)(\sigma) = i_\sigma^*(\lambda).$$

To check that $\alpha(\lambda)$ is equivariant we compute

$$\alpha(\lambda)(g\sigma) = i_{g\sigma}^*(\lambda) = (i_\sigma \hat{g})^*(\lambda)$$
$$= \hat{g}^* i_\sigma^*(\lambda) = \hat{g}^*(\alpha(\lambda)(\sigma))$$

as was to be shown. (See Chapter I, sections 5 and 6.)

We must check that α is an isomorphism. We shall show that its inverse is given by the map

$$\beta : C_G^p(K, h^q) \to \tilde{\mathcal{H}}^q(C_p^+)$$

defined as follows: Let $f \in C_G^p(K, h^q)$. Note that

$$j_{g\sigma}^*(f(g\sigma)) = (\hat{g}^{-1} j_\sigma)^*(\hat{g}^*(f(\sigma))) = j_\sigma^*(f(\sigma)).$$

Let $T \subset C_p$ be a system of representatives of the orbits of G on the set C_p and define

$$\beta(f) = \prod_{\sigma\in T} j_\sigma^*(f(\sigma)). \tag{4.4}$$

Now we compute

$$\alpha(\beta(f))(\sigma) = i_\sigma^*(\beta(f)) = i_\sigma^*\Big(\prod_{\tau\in T} j_\tau^*(f(\tau))\Big)$$
$$= i_\sigma^* j_\sigma^*(f(\sigma)) = (j_\sigma i_\sigma)^*(f(\sigma)) = f(\sigma)$$

so that $\alpha\beta = 1$. Also

$$\beta(\alpha(\lambda)) = \prod_{\sigma\in T} j_\sigma^*(\alpha(\lambda)(\sigma))$$
$$= \prod_{\sigma\in T} j_\sigma^*(i_\sigma^*(\lambda)) = \prod_{\sigma\in T} (i_\sigma j_\sigma)^*(\lambda) = \lambda$$

so that $\beta\alpha = 1$. Thus α is an isomorphism as was to be shown.

Now we claim that under the isomorphism

$$E_1^{p,q} \approx \tilde{\mathscr{H}}^q(C_p^+) \xrightarrow{\alpha} C_G^p(K;h^q)$$

the differential d_1 becomes, up to sign, the coboundary.

We first remark that, up to sign, $d_1\colon E_1^{p,q} \to E_1^{p+1,q}$ may be identified with the homomorphism

$$\tilde{\mathscr{H}}^{p+q}(K^p/K^{p-1}) \xrightarrow{\approx} \tilde{\mathscr{H}}^{p+q+1}(S(K^p/K^{p-1})) \xrightarrow{\psi_p^*} \tilde{\mathscr{H}}^{p+q+1}(K^{p+1}/K^p)$$

where $\psi_p\colon K^{p+1}/K^p \to S(K^p/K^{p-1})$ is an equivariant map defined as follows: If σ is a $(p+1)$-cell and $f_\sigma\colon S^p \to K^p$ is a characteristic map (chosen equivariantly) we follow f_σ by collapsing $K^p \to K^p/K^{p-1}$ and suspending $S^{p+1} \to S(K^p/K^{p-1})$ (unreduced on the left, reduced on the right). Then the cell $\sigma/\dot{\sigma} \subset K^{p+1}/K^p$ is identified with S^{p+1} in a canonical way (taking the base point into the north pole of S^{p+1}). The resulting maps $\sigma/\dot{\sigma} \to S(K^p/K^{p-1})$ are put together to form the map $\psi_p\colon K^{p+1}/K^p \to S(K^p/K^{p-1})$. The verification of this relies on the fact that in the Puppe sequence for the inclusion $i\colon K^p/K^{p-1} \to K^{p+1}/K^{p-1}$ the map $C_i \to S(K^p/K^{p-1})$ may be identified with ψ_{p+1}. The details will be left to the reader.

Now $K^{p+1}/K^p \approx S^{p+1}C^+_{p+1}$ and $S(K^p/K^{p-1}) \approx S^{p+1}C^+_p$ so that the map ψ_p is described by the induced maps $\sigma/\dot\sigma \subset S^{p+1}C^+_p \to S^{p+1}C^+_p = \bigvee_\tau S(\tau/\dot\tau) \to S(\tau/\dot\tau)$ (where $\sigma \varepsilon C_{p+1}, \tau \varepsilon C_p$). It is easy to see that, in fact, this map has degree $[\tau\colon \sigma]$ (see Chapter I, section 1).

Thus d_1 is induced, up to sign, by

$$\eta_p^*\colon \tilde{\mathcal{H}}^{q+1}(SC^+_p) \to \tilde{\mathcal{H}}^{q+1}(SC^+_{p+1})$$

where $\eta_p\colon SC^+_{p+1} = \bigvee_\sigma S_\sigma \to \bigvee_\tau S_\tau = SC^+_p$ is an equivariant map such that the induced map $S_\sigma \to S_\tau$ has degree $[\tau\colon \sigma]$. (Here we use S_σ to stand for a copy of the circle indexed by the cell σ.)

We claim that the following diagram commutes

$$\begin{array}{ccc} \tilde{\mathcal{H}}^{q+1}(SC^+_p) & \xrightarrow{\eta_p^*} & \tilde{\mathcal{H}}^{q+1}(SC^+_{p+1}) \\ \downarrow{\scriptstyle \alpha S^{-1}} & & \downarrow{\scriptstyle \alpha S^{-1}} \\ C^p_G(K;h^q) & \xrightarrow{\delta^p} & C^{p+1}_G(K;h^q) \end{array} \tag{4.5}$$

where we use S to denote the suspension isomorphism. The proof is straightforward but will involve some cumbersome details. First, suppose σ is a $(p+1)$-cell and τ is a p-cell of K with $K(\tau) \subset K(\sigma)$. Then let θ^τ_σ denote the equivariant map $G/G_\sigma \to G/G_\tau$ induced by inclusion $G_\sigma \subset G_\tau$. Using $[\tau\colon \sigma]$ to denote maps of <u>degree</u> $[\tau\colon \sigma]$ we note that the diagram

$$\begin{array}{ccccc} S(G/G_\sigma)^+ & \xrightarrow{Si_\sigma} & SC^+_{p+1} & \xrightarrow{\eta_p} & SC^+_p \\ \downarrow{\scriptstyle \bigvee[\tau:\,\sigma]} & & & & \uparrow{\scriptstyle \bigvee Si_\tau} \\ \bigvee_{\tau\varepsilon T} S(G/G_\sigma)^+ & & \xrightarrow{\bigvee S\theta^\tau_\sigma} & & \bigvee_{\tau\varepsilon T} S(G/G_\tau)^+ \end{array}$$

of equivariant maps commutes, where T is the set of all p-cells τ with $K(\tau) \subset K(\sigma)$.

The induced diagram in cohomology is

$$\begin{array}{ccccc}
\tilde{\mathcal{H}}(S(G/G_\sigma)^+) & \xleftarrow{(Si_\sigma)^*} & \tilde{\mathcal{H}}(SC^+_{p+1}) & \xleftarrow{\eta_p^*} & \tilde{\mathcal{H}}(SC^+_p) \\
{\scriptstyle \sum_\tau [\tau:\,\sigma]} \uparrow & & & & \downarrow {\scriptstyle \sum_\tau (Si_\tau)^*} \\
\sum_{\tau \in T} \tilde{\mathcal{H}}(S(G/G_\sigma)^+) & & \xleftarrow{\Sigma (S\theta^\tau_\sigma)^*} & & \sum_{\tau \in T} \tilde{\mathcal{H}}(S(G/G_\tau)^+)
\end{array}$$

Since $(S\varphi)^* = S \circ \varphi^* \circ S^{-1}$ we obtain from this diagram that

$$(4.6) \qquad (Si_\sigma)^* \eta_p^* = S[\sum_\tau [\tau:\,\sigma](i_\tau \theta^\tau_\sigma)^*]S^{-1}.$$

Now let us verify that (4.5) commutes. Let $\lambda \varepsilon \tilde{\mathcal{H}}^{q+1}(SC^+_p)$ and, as usual, let σ be a (p+1)-cell of K. Then

$$\begin{aligned}
\alpha(S^{-1}(\eta_p^*(\lambda)))(\sigma) &= i_\sigma^*(S^{-1}(\eta_p^*(\lambda))) \\
(4.7) \qquad &= S^{-1}(Si_\sigma)^* \eta_p^*(\lambda) \\
&= \sum_\tau [\tau:\,\sigma](i_\tau \theta^\tau_\sigma)^*(S^{-1}\lambda).
\end{aligned}$$

(The last equality comes from (4.6).) On the other hand

$$\delta^p(\alpha S^{-1}(\lambda))(\sigma) = \sum_\tau [\tau:\,\sigma](\theta^\tau_\sigma)^*(\alpha(S^{-1}(\lambda))(\tau))$$

directly from the definition of δ^p. This may be further simplified to

$$\sum_\tau [\tau:\,\sigma](\theta^\tau_\sigma)^* i_\tau^*(S^{-1}\lambda),$$

the same as in (4.7). This shows that (4.5) commutes and hence, finally, that $d_1\colon E_1^{p,q} \to E_1^{p+1,q}$ becomes the coboundary under our isomorphism with $C_G^*(K;h^q)$. Thus we have

$$(4.8) \qquad E_2^{p,q} \approx H_G^p(K;h^q).$$

As noted before, the spectral sequence converges (when $\dim K < \infty$) to the graded group associated with some filtration of $\mathcal{H}^{p+q}(K)$.

5. The "classical" uniqueness theorem

Suppose that $\mathcal{H}^*$ is an equivariant cohomology theory satisfying the dimension axiom (4) of section 2, Chapter I. Let $h \in \mathcal{C}_G$ denote the "coefficients" of this theory. That is $h(G/H) = \mathcal{H}^0(G/H)$, and so on. Let K be a finite dimensional G-complex. If K is infinite we assume that (A) of the last section is satisfied.

In this case the spectral sequence of the last section degenerates for $r \geq 2$. In fact

$$E_2^{p,q} = \begin{cases} H_G^p(K;h); & q = 0 \\ 0 & ; q \neq 0 \end{cases}$$

It follows that, in fact,

$$\mathcal{H}^p(K) \approx H_G^p(K;h) \tag{5.1}$$

and naturality is not hard to verify. Thus this is the <u>only</u> equivariant classical cohomology theory having coefficients h. The reader should note that, for general $h \in \mathcal{C}_G$, h is indeed the coefficient system of the cohomology theory $H_G^*(K;h)$. That is, there is a natural isomorphism

$$h(G/H) \approx H_G^0(G/H;h).$$

Lecture Notes in Mathematics

Bisher erschienen/Already published

Vol. 1: J. Wermer, Seminar über Funktionen-Algebren. IV, 30 Seiten. 1964. DM 3,80

Vol. 2: A. Borel, Cohomologie des espaces localement compacts d'après J. Leray. IV, 93 pages. 1964. DM 9,–

Vol. 3: J. F. Adams, Stable Homotopy Theory. 2nd. revised edition. IV, 78 pages. 1966. DM 7,80

Vol. 4: M. Arkowitz and C. R. Curjel, Groups of Homotopy Classes. Second revised edition. IV, 36 pages. 1967. DM 4,80

Vol. 5: J.-P. Serre, Cohomologie Galoisienne. Troisième édition. VIII, 214 pages. 1965. DM 18,–

Vol. 6: H. Hermes, Eine Termlogik mit Auswahloperator. IV, 42 Seiten. 1965. DM 5,80

Vol. 7: Ph. Tondeur, Introduction to Lie Groups and Transformation Groups. VIII, 176 pages. 1965. DM 13,50

Vol. 8: G. Fichera, Linear Elliptic Differential Systems and Eigenvalue Problems. IV, 176 pages. 1965. DM 13.50

Vol. 9: P. L. Ivănescu, Pseudo-Boolean Programming and Applications. IV, 50 pages. 1965. DM 4,80

Vol. 10: H. Lüneburg, Die Suzukigruppen und ihre Geometrien. VI, 111 Seiten. 1965. DM 8,–

Vol. 11: J.-P. Serre, Algèbre Locale. Multiplicités. Rédigé par P. Gabriel. Seconde édition. VIII, 192 pages. 1965. DM 12,–

Vol. 12: A. Dold, Halbexakte Homotopiefunktoren. II, 157 Seiten. 1966. DM 12,–

Vol. 13: E. Thomas, Seminar on Fiber Spaces. IV, 45 pages. 1966. DM 4,80

Vol. 14: H. Werner, Vorlesung über Approximationstheorie. IV, 184 Seiten und 12 Seiten Anhang. 1966. DM 14,–

Vol. 15: F. Oort, Commutative Group Schemes. VI, 133 pages. 1966. DM 9,80

Vol. 16: J. Pfanzagl and W. Pierlo, Compact Systems of Sets. IV, 48 pages. 1966. DM 5,80

Vol. 17: C. Müller, Spherical Harmonics. IV, 46 pages. 1966. DM 5,–

Vol 18: H.-B. Brinkmann und D. Puppe, Kategorien und Funktoren. XII, 107 Seiten. 1966. DM 8,–

Vol. 19: G. Stolzenberg, Volumes, Limits and Extensions of Analytic Varieties. IV, 45 pages. 1966. DM 5,40

Vol. 20: R. Hartshorne, Residues and Duality. VIII, 423 pages. 1966. DM 20,–

Vol. 21: Seminar on Complex Multiplication. By A. Borel, S. Chowla, C. S. Herz, K. Iwasawa, J.-P. Serre. IV, 102 pages. 1966. DM 8,–

Vol. 22: H. Bauer, Harmonische Räume und ihre Potentialtheorie. IV, 175 Seiten. 1966. DM 14,–

Vol. 23: P. L. Ivănescu and S. Rudeanu, Pseudo-Boolean Methods for Bivalent Programming. 120 pages. 1966. DM 10,–

Vol. 24: J. Lambek, Completions of Categories. IV, 69 pages. 1966. DM 6,80

Vol. 25: R. Narasimhan, Introduction to the Theory of Analytic Spaces. IV, 143 pages. 1966. DM 10,–

Vol. 26: P.-A. Meyer, Processus de Markov. IV, 190 pages. 1967. DM 15,–

Vol. 27: H. P. Künzi und S. T. Tan, Lineare Optimierung großer Systeme. VI, 121 Seiten. 1966. DM 12,–

Vol. 28: P. E. Conner and E. E. Floyd, The Relation of Cobordism to K-Theories. VIII, 112 pages. 1966. DM 9,80

Vol. 29: K. Chandrasekharan, Einführung in die Analytische Zahlentheorie. VI, 199 Seiten. 1966. DM 16,80

Vol. 30: A. Frölicher and W. Bucher, Calculus in Vector Spaces without Norm. X, 146 pages. 1966. DM 12,–

Vol. 31: Symposium on Probability Methods in Analysis. Chairman: D. A. Kappos. IV, 329 pages. 1967. DM 20,–

Vol. 32: M. André, Méthode Simpliciale en Algèbre Homologique et Algèbre Commutative. IV, 122 pages. 1967. DM 12,–

Vol. 33: G. I. Targonski, Seminar on Functional Operators and Equations. Approx. IV, 110 pages. 1967. DM 10,–